Materials for Automobile Bodies

DATE DUE

Materials for Automobile Bodies

Geoff Davies F.I.M., M.Sc. (Oxon)

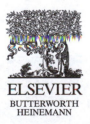

ELSEVIER
BUTTERWORTH
HEINEMANN

AMSTERDAM   BOSTON   HEIDELBERG   LONDON   NEW YORK   OXFORD
PARIS   SAN DIEGO   SAN FRANCISCO   SINGAPORE   SYDNEY   TOKYO

Butterworth-Heinemann
An imprint of Elsevier
Linacre House, Jordan Hill, Oxford OX2 8DP
200 Wheeler Road, Burlington MA 01803

First published 2003

British Library Cataloguing in Publication Data
A catalogue record for this book is available from the British Library

Library of Congress Cataloguing in Publication Data
A catalogue record for this book is available from the Library of Congress

ISBN   0 7506 5692 1

For information on all Butterworth-Heinemann
publications visit our website at www.bh.com

Typeset by Replika Press Pvt Ltd, India
Printed and bound in Great Britain by Biddles Ltd, *www.biddles.co.uk*

Contents

Preface

The field of materials used for automotive body construction has become increasingly complex over the last 10 years or so, whilst time available for the understanding of the various technologies involved has diminished. As well as body specialists the current volume will be additionally useful to those in peripheral areas of automotive engineering and associated manufacturing and purchasing functions, aware of much of the terminology used but who now wish to understand specific elements of materials technology in greater breadth and depth. This would include processing aspects such as hydroforming, laser welding, tailor welded blanks and polymeric materials, as well as steel and aluminium, and hopefully this will assist them in their decision making where choices in design or sourcing have to be made. Those in the automotive supply industry should also find it instructive in understanding the objectives and requirements of design engineers, and the processing expectations of technologists involved with component manufacture ('the process chain' or 'AMC', the automotive process chain). Above all, it is hoped that students and future generations of automotive design and related engineers should benefit from an introduction to all topics relevant for the understanding of automotive body materials technology, now gathered for once in a single digestible volume.

It is intended that the format, starting with the initial specification of the material and concluding with the issues involved in the final vehicle disposal, is straightforward and presents the facts he/she 'needs to know', with the presentation of detail only when it may be justified. However, should additional information be required, authoritative references are provided.

As well as conveying the properties relevant to the selection and use of body materials, together with essential features of their preparation, two recurrent themes are emphasized. The first is the need for material consistency which is a major factor in achieving maximum utilization in production and the second is the need for thorough rehearsal of any new aspect of materials technology before implementation is even considered. The evolutionary principle whereby new technology is gradually assessed and introduced in low volume applications, before introduction to high production models is a major feature of all successful European and Japanese marques. The urge to release innovatory technology too early – however attractive potentially – can often cause pain, in the form of delayed launch time and wasted resources.

With the advent of computer aided design procedures it has become increasingly incumbent on the structural engineer himself to make a material choice from a number of predetermined choices programmed into his particular system. The opportunity may not nowadays always exist to consult a materials engineer who used to be steeped in many decades of experience in evolving designs and appreciated the many hidden issues that can arise during processing and service. The 'obvious' choice may not be too obvious and involve indirect costs and consequences which may unexpectedly occur at a future date. The need exists therefore for easily digested explanations of materials 'pros and cons' to help him in his choice and assess implications of associated treatments which again may not be immediately apparent to the non-specialist. Likewise it may also assist the purchasing agent in understanding price extras applying to many apparently similar materials options, and treatments, plus other specialists and suppliers involved in the 'automotive process chain' seeking to anticipate the effects of changing specifications on their processes. Even to the more experienced design engineer it will be obvious that the field of associated body materials technology has become increasingly complex over the past twenty years and a reference to the 1970s will quickly show that the selection of materials was then limited to a simple selection of 4–5 sheet forming grades. Processing technology has changed significantly and the introduction of a comparatively small number of changes in steel and aluminium grades and choice of several coating types have served to generate a surprising level of confusion. Thus for even the seasoned engineering specialist the need also exists to augment existing knowledge and it is hoped that this volume will fulfil this need or at least provide a pointer as to where that data can be found.

Wherever possible examples of production bodies have been used to illustrate the development of the various design technologies and the experience of key manufacturers such as BMW (body design/ environmental aspects) and Volvo (safety engineering) employed to illustrate the synergies achieved by the use of newer materials. This experience is augmented by the massive resources that have been utilized within the ULSAB programmes by the steel industry and also draws on the ASV(Alcan) and ASF(Audi) technology to demonstrate the potential of aluminium.

Geoff Davies

Acknowledgements

The author would like to acknowledge the contributions made by a host of colleagues in the automotive materials and associated supply fields, and academia. Particular thanks are due to Dr Richard Holliday for his initial encouragement and authoritative chapter on 'Joining', together with background preparation for other chapters including the polymer content. Many thanks are due to Dr John Sykes, University of Oxford, and Dr David Worsley of the University of Swansea for their help in preparing the chapter on 'Corrosion'. Also to Brian Simonds for inserting the original artwork and Mike Boyles and his team at Corus for proof reading sections on the various aspects of steel technology. Similar thanks are due to Roy Woodward for his helpful guidance with respect to the latest aluminium technology. The contribution of Brian O'Rourke on Formula 1 and competition body materials was invaluable, and because of its comprehensive yet concise insight into the complex and wider issues surrounding the choice of F1 composite materials his contribution, originally made to a companion volume, has been updated and reproduced in its entirety. Finally, special thanks are due to the author's family for their patience and encouragement over many months.

Acknowledgements

About the author

Geoff Davies began his career in the materials field within the Richard Thomas and Baldwins Foundry and Refractories plant at Landore near Swansea in 1956, later moving to the R.T.B. Central Research Laboratories at Whitchurch near Aylesbury in Buckinghamshire where his interest in the production technology of steel sheet and coated flat strip began. Development work included low carbon sheet steels with enhanced drawability, lubricity studies during sheet forming, aluminium clad steel produced by roll-bonding, and optimization of various forms of hot-dip galvanized coatings including differential and iron–zinc alloy types. At this stage he had obtained his Associateship of the Institution of Metallurgists examination by part-time study at the Sir John Cass College in London.

Following ten years on steel production and supply technology this experience proved invaluable when the author then moved to the R and D Department of the Pressed-Steel Company at Cowley, Oxford, to join the team working on the formability of sheet steel for automotive bodies, led by Dr John Wallace and Dr Roger Butler who had strong ties with the University of Birmingham sheet forming experts led by Prof. D.V. Wilson. Thirty-five further years with this and associated organizations, which successively included the Manufacturing Technology Department of Austin-Rover, BL Technology at Gaydon, returning to Rover at Oxford and finally BMW UK, completed 45 years of experience on the application of mainly – but not exclusively – ferrous materials, chiefly within the motor car body. To complete his qualifications he was elected a Fellow of the Institute of Materials in 1994 and was awarded an M.Sc. degree at Keble College Oxford in 1999 following research work on the electrochemistry of zinc coated steels in automotive applications with Dr John Sykes.

Since 1980 he has represented SMMT on BSI Committee ISE 10 providing the automotive input on flat rolled steel products, and commencing in 1971 has written over 20 papers on the subject in major technical journals. Being chairman of the SMMT Steel Sheet Rationalization group has enabled a balanced view to be presented reflecting input from most of the major UK motor manufacturers.

Participation in major EC-sponsored research programmes such as BRITE-EURAM Low Weight Vehicle and Alternative Zinc Coating projects and the ECSC Adhesively Bonded Structures initiative has ensured an awareness of most recent developments in the field of automotive body materials. Close co-operation with Corus and other steel manufacturing technical operations has ensured that maximum use has been made of high strength steel and coated sheet innovations over the last 30 years to ensure that weight reduction targets have progressively been met together with a major improvement in vehicle corrosion resistance and overall steel quality. Similar ties have been developed with major aluminium producers such as Alcan International over many years to ensure maximum use has been made of improvements in alloy and surface technology by Land Rover, one of the most prominent aluminium users over the years. Membership of a DTI Technical Mission to Japan on laser application in industry proved a valuable insight into the possibilities of this and associated automation techniques in automotive body assembly.

Disclaimer

The author and publisher would like to thank those who have kindly permitted the use of images in the illustration of this book. Attempts have been made to locate all the sources of illustrations to obtain full reproduction rights, but in the very few instances where this process has failed to find the copyright holder, apologies are offered. In the case of an error, correction would be welcomed.

1 Introduction

1.1 Overview of content

The purpose of this textbook is to provide an easily understood presentation of the technology surrounding the choice and application of the main materials that can be used for the construction of the automotive body structure. Although many reference works exist on specific designs and associated materials, in the form of books or conference proceedings, these tend to focus on individual materials, test methods or numerical simulations. Relatively few have attempted to appraise all realistic candidate materials with regard to design, manufacture, suitability for component production, corrosion resistance and environmental attributes, against relevant selection criteria, all within a single volume. The problem with such a comprehensive presentation is in limiting the factual data on each subject, and it is hoped that the content is presented in sufficient depth to enable an understanding of the relevance of each topic, without overpowering the non-specialist with too much detail.

1.2 Materials overview

Although the bodywork conveys the essential identity and aesthetic appeal of the vehicle, typified by the styling panache of Aston Martin or Ferrari, or the drab functionality of utility vehicles such as the Trabant, the actual material from which it is fabricated has until recently attracted relatively little interest. However, of all the components comprising the overall vehicle, the skin and underlying structural framework provide some of the most interesting advances in materials and associated process technology. This is reflected in the many changes which have taken place in the body materials used for the automotive body structures over 100 years of production, commencing with the replacement of the largely handcrafted bodies constructed from sheet metal, fabric and timber superseded during the 1920s, by sheet steel. Low carbon mild steel strip was favoured by concerns such as the Budd Company and Ford in the USA, due to the faster production rates attainable by press forming of panels from flat blanks and subsequent assembly using resistance welding techniques. This trend to mass production quickly established itself in Europe and elsewhere, through offshoots of these companies such as the Pressed Steel Company at Oxford in the UK, and the Ford Dagenham plant, and steel has remained the predominant material ever since.[1] Therefore for any introductory text on this subject it is fair that the initial emphasis should focus on steel and its variants, and the technology associated with its use. Aluminium has long been recognized as a lightweight alternative although economics have made it the second choice for the body architecture of models to be produced in any numbers. However, as other criteria, such as emissions control now become increasingly prominent, its potential for energy efficient mass production vehicles in the future has once again to be acknowledged, as already demonstrated by recent models such as the Audi A2, now available in Europe, and this is also reflected in the wide coverage given in the following chapters. The same applies to other materials such as magnesium where wider application is now being considered. Although plastics also provide a lighterweight alternative to steel and provide greater freedom of exterior styling, some conflict now arises with another environmental requirement imposed by recycling targets with a large proportion of plastics still reliant on landfill sites for disposal. The materials selection procedure adopted by most major 'environmentally friendly' manufacturers recognizes an increasing range of requirements, now extended to include 'process chain' compatibility, i.e. ease of application within the manufacturing cycle, together with the need to consider the total life cycle of materials used (with respect to factors such as cost, energy and disposal). This is illustrated in simple tabular form within the opening section to Chapter 3, which summarizes realistic choices viewed against engineering and the other key criteria already mentioned.

Steel has demonstrated an all-round versatility over many years. It has remained reasonable in cost, life of pressed components has been extended through the use of zinc coating technology, and the range of strength levels has increased to meet increasingly stringent engineering needs, and very importantly, but often overlooked, it is very adaptable with regard to corrective rework. This may be required on-line, to rectify production defects which can sometimes occur with the best of manufacturing systems, or for repair purposes following accidental damage in service, but experience has shown steel is highly tolerant to reshaping and a large infrastructure exists of skills and material to restore the structure to meet the original engineering specification. Often disregarded, ease and cost of repair are now increasingly important as newer grades of high strength steels and aluminium and other materials emerge, requiring precise retreatments involving more sophisticated equipment to ensure original standards can be met. These procedures can have an important bearing on the insurance category derived for specific vehicles.

A few introductory facts might help at this stage, prior to detailed discussion later. Unless otherwise stated the main discussion centres on the sheet condition although the importance of tubular construction and other material forms will become evident in later sections. Depending on grade and precise model design specification, approximately half a tonne of steel strip is required to produce a body of unitary construction and between 40 and 45 per cent of this is discarded in the form of press-shop scrap. This may arise through unavoidable areas of the blank which cannot be utilized due to mismatch of shape with strip dimensions, or otherwise by non-productive press strokes. Currently one tonne of prime steel costs around £360, according to the specification ordered, and although most of the scrap is recyclable, the value for baled offal is only approximately one-eighth of the original price. According to Ludke[2] the expenditure on body materials accounts for approximately 50 per cent of the B.I.W. costs.

Steel thickness as indicated by external panels has shown an overall reduction over the years from 0.9 mm in the 1930s to the current norm of 0.75 mm, due mainly to pressures to improve cost savings through increased yield for successive models, and more recently the emergence of dent resistant grades ensuring less cosmetic damage at thinner gauges in service. Similar trends have been noted for internal parts where stiffness (a basic design parameter) is not compromised, but from a manufacturing standpoint, as noted previously[3] the key requirements have remained for weldable grades which could be formed with the least possible expense.

Historically, these normally fell into flat, deep drawing or extra deep drawing qualities, were either rimmed or killed (stabilized), made by the ingot cast production rate, and, apart from rare instances, up until the 1960s showed a yield strength of 140 MPa. As explained later, the ingot casting route has now been displaced by the more consistent and economic continuous casting process. Although the strength did not at this stage vary greatly, many improvements were taking place in the drawing properties, surface technology and consistency of the products, and it is important to briefly recap the historical detail illustrating the development of these properties and interrelationship with processes during the course of the book.

It was during the 1980s that more significant influences began to emerge, beginning with the increasing use of zinc coated steels. While volume production was the key priority of the 1960s it is probably fair to say that bodies produced during this period were vulnerable to corrosion. This was a consequence of economics (higher yield in blanks per tonne) dictating thinner gauges and the demands for higher volume production which often called for shorter cycle times in the paint process and consequent risk of incomplete coverage. It wasn't until the 1970s that the more efficient cathodic electropriming painting systems were developed and galvanized steel gradually introduced. It is interesting to note that hints were being dropped in the corrosion repair manuals of the day[4] that poor longevity was partly attributable to built-in obsolescence and that the use of zinc coated steels would provide an answer. While it was possible to see galvanized panels as the simple solution, in reality it required 20 years of steady development jointly between steel suppliers and car manufacturers between 1960 and 1980 to ensure that a consistent product could be adapted to the demands of automation in BIW assembly, while achieving the ever increasing standards of paint finish required by the consumer. Enormous strides have been made in the protection of the car body over the last 30 years and this is reflected in the design targets of most manufacturers which has advanced to 12 years (in some instances 30 years) of freedom from perforation.

In considering longevity it is necessary to draw the distinction between the emphasis on the materials and engineering senses of 'durability', and it will be apparent that the focus here is on corrosion mechanisms and modes of protection rather than physical and mechanical endurance aspects of vehicle life which are perhaps outside the scope of this work. At the outset it is emphasized that this volume is not meant to offer any instruction for repair of corroded examples of materials covered here, but it will be obvious that the materials utilized are increasingly specialized. As stated for higher strength grades above, it is essential therefore for the preservation of optimum strength levels and corrosion resistance that repair techniques are constantly updated, and recommended procedures amended to maintain engineering properties and vehicle life. The need to heed manufacturers' recommendations is paramount if safety standards are to be maintained.

During this period other strong influences were emerging calling for increased safety standards with regard to occupant and pedestrian protection and again the relevant background and design aspects to developments will be discussed, although the main theme will be on the synergies and improvements achieved through the utilization of materials. Coincidentally the first legislation was being called for in the USA requiring improved fuel consumption partly as a result of the 1973 world oil crisis and also to pacify a growing anti-pollution lobby. This resulted in Corporate Average Fuel Economy (CAFÉ) regulations which specified average fleet targets (e.g. 27.5 mpg for 1985) which all motor organizations trading in America had to meet. The response from vehicle manufacturers included the use of thinner gauge steel to lighten structures, utilizing initially high strength low alloy steels and then rephosphorized grades. For the body shell these were used for applications such as longitudinal members and 'B' posts where use could be made of increased energy absorption. It was also found that door panel thickness could be reduced using bake-hardening steels which strengthened in the paint ovens by strain ageing and enabled downgauging on the basis of higher dent resistance. The increased utilization of these steels through various generations of vehicles is shown in Fig. 1.1.

Latterly environmental concerns, which were emerging through the 1980s and 1990s on issues such as emissions control, have heightened the need for weight reduction, and governmental pressures have appeared requiring progressively lower levels of CO_2 in terms of grams per kilometre, with tax incentives for lower rated cars. This

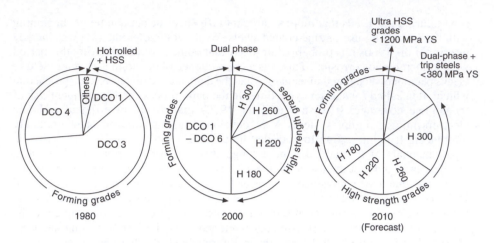

Fig. 1.1 Increasing utilization of high strength steel grades

has prompted the need for even lighter structures and has strengthened the case for aluminium and plastics. However, as mentioned above, in the case of polymeric materials some conflict now arises with regard to weight reduction, as further environmental pressures and legislation concerning recycling and vehicle disposal, accompanied by governmental targets, requires improvements in the amount of material recycled within a specific timeframe. Currently 75 per cent of the vehicle is reprocessed – the remaining 'fluff' being essentially non-metallic and comprising a high polymeric content. The background and tasks that lie ahead for the material specialists concerning identification for disassembly, preferred types and reuse of materials are outlined later, together with progress achieved worldwide and possible future solutions. Practices already adopted by more progressive motor manufacturers are also highlighted.

Aluminium has always been considered an alternative material to steel and instances can be recalled of it being used for models in the early 1900s and volume production in the 1950s, but until recently economics, both initial material and processing costs, have discouraged widespread adoption. A general rule[5] was that the net cost of aluminium assemblies after allowances for density, equivalent section, modified processing, etc., was double that of steel although weight could be halved. This has been broadly proven and vehicle structures have demonstrated that compared with the 25 per cent weight reduction achievable with steel, savings of nearer 50 per cent can be obtained with aluminium. In some designs significant changes have been made to spaceframe architecture, incorporating castings and extrusions as well as sheet, but this has given significant improvements in strength, assembly and ease of repair, and as in the case of the Audi A2 a spaceframe that can be easily mass produced. The trend in skin panel thickness in aluminium has also changed with 1.0 mm being the current norm, as opposed to 1.4–1.6 mm used previously (although mainly for luxury cars). Alloy types have changed over the last 40 years from the wrought Al–Mg alloys of the 5xxx series (susceptible to 'stretcher-strain' markings) to the heat-treatable 6xxx Al–MgSi.

Polymeric materials have massive advantages in terms of panel complexity that can be achieved in one operation and also the ease of incorporation of many parts in one assembly such as complete front ends. 'User friendliness' is also claimed in terms of low speed impact damage, e.g. gate-post scuffing, and pedestrian contact. However, incorporating the large numbers of different blends that can exist and accommodating these within the manufacturing 'process chain', i.e. manufacturing sequence, where fixing and temperature incompatibilities can exist, remain a problem. Polycarbonate, Noryl GTX, and carbon fibre composites are all examples of the range of materials in use today, and even this range illustrates the need for rationalization of different types that must be made if recycling and reuse are to be made more universal.

Magnesium is also finding limited application within the structure and although this has been an option for items such as gear box covers in the past, corrosion and ease of forming have constrained the scope of application. With improved alloys, proven in the aero industry, the time has come for reappraisal and at least one major company has followed Fiat in adopting magnesium for the fascia crossbeam.

The importance of the 'process chain' to the automotive manufacturer in the selection of materials cannot be overemphasized. The physical and mechanical properties are of basic engineering importance but unless the material can be accommodated comfortably within the operating parameters governing the sequence of manufacturing operations it will have adverse effects on both productivity and facility costs. As stated above the scope of the normal criteria must now also be widened to recognize environmental factors and controls which are becoming increasingly evident/restrictive, together with unit material cost. Therefore before more detailed presentations of each of the main material contenders in Chapter 3, a summary table is constructed which is intended to prompt a wider perspective for the decision-making process on the material choice than that based purely on inherent properties. Using factors which include suitability for various aspects of manufacture and likely environmental implications, the effects of various choices on facility investment or meeting emissions or disposal regulations can be weighed more comprehensively. The ratings used are not definitive and must depend on the constructors' circumstances and information available at the appropriate time, but this table does illustrate the wider approach now necessary in the selection rationale. A tabular guide as to materials specified under more defined conditions of volume and implementation time ('normal' or 'accelerated') is also proposed in Chapter 9 based on types of technology already evident in production or for which production feasibility has been proven.

1.3 General format of presentation

The layout of the book follows a logical sequence beginning with a consideration of the design configuration and how this has evolved through many distinct phases, and demonstrating how best use has been made of the material characteristics and format selected, and the synergies derived. This brief historic presentation shows how various configurations have evolved from separate chassis and body to the latest spaceframe. Key 'milestones' in body design are identified and it is demonstrated how one major manufacturer has exploited material properties through three generations of similar models.

It is then appropriate that characteristics of the materials, 'the building blocks', used in the design are next introduced in detail, together with their manufacturing history, so that their capability and limitations can be recognized in terms of fitting engineering, product and process profiles. As well as the widening choice of high strength and zinc coated steels, fundamental changes have also been introduced in the form of steels and other strip materials that can now be utilized for localized strengthening and parts consolidation. Tailor welded blanks offer the design engineer localized strengthening by the use of differential strength/thickness/coating combinations and allowing specific engineering requirements to be used exactly where needed.

The same applies to hydroformed sections and although the associated joining issues are more complex, formed tubular sections may be used to achieve significant parts rationalization. Again this represents a step change in material utilization and examples will be given of the assessment of this type of technology applied to existing models within the section on component manufacture. Figure 1.2 shows the extent to which these various forms could be utilized as demonstrated by the recent ULSAB programme sponsored by a group of 35 worldwide steel producers. Techniques such as laser welding have also seen recent application allowing stiffening of sections during assembly, enhancing the properties and even allowing localized heat treatment. An important design tool for engineering more rigidity into roof/cantrail sections, for example, is included in the chapter on Joining.

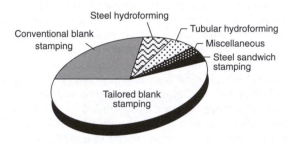

Fig. 1.2 Utilization of different material forms within future body structures. (Courtesy of ULSAB Consortium)

Ease of component manufacture and implications regarding the accommodation of materials within the 'process chain' are of vital importance in the assessment of effects on existing facilities and identification of future requirements, and this is covered in detail in the chapters on 'Component Manufacture' and 'Joining'.

To demonstrate how innovative ideas are evaluated and proven for new designs it was also felt that several broader initiatives and development programmes should be highlighted. These have not been necessarily adopted for production cars, but have provided invaluable feedback to niche and volume car designers, and, deserving of a collective chapter, this is presented under the title of 'Demonstration, Concept and Competition Cars'. This includes wide ranging steel and aluminium industry projects, European programmes and reference to futuristic visualizations such as the 'Hypercar'. The use of concept cars in judging public acceptance of new ideas is discussed with several different innovations highlighted, and a fascinating insight is also provided into materials used for Williams Formula 1 racing cars.

As stated, corrosion resistance is obviously of extreme importance in the design of the body structure and all aspects including preferred panel design principles, processing, vehicle assessment and simulated test methods are included on this subject, which is considered a key issue of consumer concern. Within the industry it is also an area of extreme competitor awareness and therefore we have accorded it a separate chapter.

Because of their influences on the well-being of the individual a separate chapter has been created on materials and their influence on improved vehicle safety, emissions and recycling ELV issues. As this illustrates the positive attitude adopted by the manufacturer in these areas of human concern this is entitled 'Environmental Considerations'.

Finally, the future is discussed and material options for directions for body designs debated considering various timeframes and circumstances. A key issue appears to be the type of fuel system to be adopted, as the status of such systems strongly influences the weight requirements of the supporting structure. Alternatives to current means of propulsion appear to be on the horizon and if these are thrust on us sooner than later there could be a need for radical changes quickly, as ultralightweight structures are suddenly required to support heavyweight electric/fuel cell modes of propulsion.

Despite fuel consumption being a main concern since the 1970s when legislation in the USA dictated that specific corporate mpg levels were reached by companies distributing vehicles in that country, and more globally thereafter as pressure has mounted for emissions control, there has been very little commitment on behalf of the consumer over the last 30 years. 'Gas guzzlers' still abound and the popularity of 4×4 vehicles appears to be growing. Companies are making real efforts to meet legislative controls by producing cars with CO_2 emissions levels reduced below 100 g/km, but is there a real commitment from the public to drive such vehicles without stronger controls being imposed? Time is required to resolve political issues, for example, in the USA, where a lobby exists which emphasizes the negative effect of downweighting on safety, and in Europe concerning the provision of an alternative fuel pump infrastructure (e.g. hydrogen), etc. Therefore combined with a notoriously conservative industry when it comes to material choice – it is possible that any change might be gradual allowing the time for development of alternative fuels to a more efficient form. In the intervening period it is also possible that solutions will be found to outstanding recycling, plastics rationalization, and processing issues. Together with further improvements in existing power train technology, hybrid structures substantially developed from today's designs should satisfy emerging requirements. To examine this situation more logically a tabular presentation is included which proposes possible material combinations that could emerge with differing circumstances, 'anticipated' and shorter-term 'accelerated' situations, within volume and niche car sectors. These are not meant to be definitive conclusions and the combinations suggested are debatable, but the reasoning is presented for each of the choices and the real purpose of this section of Chapter 9 is to stimulate balanced discussion on the various options although this may prompt other equally relevant proposals! So often the views presented at seminars and industry forums reflect vested interests or loyalties of the presenters or contributors with little time for reasoned debate. In Chapter 9 an attempt is made to weigh the pros and cons more objectively, allowing the reader to reflect at length on some realistic choices on which to base his/her vision for the future. To illustrate the choices reference is often made to tangible applications evident in current or more innovative production models.

1.4 Introduction to body architecture and terminology

At this stage some introduction to the structural elements considered by this publication is called for before the main text begins. In simple terms this covers the 'body-in-white', a traditional name for the body structure which is depicted in its component form in Fig. 1.3.

'Closures' as the drawing shows refer to all those panels or subassemblies which are attached mechanically to the main substructure by hinges or other means, and hence the alternative reference an 'bolt on' panels. Other terms or acronyms which are commonly encountered in the text include:

Fenders = wing panels

Trunk = boot lid

Rocker panels = sill sections

Deck = bonnet lid

Wheel-house = wheel arch inner

Fig. 1.3 Components subassemblies comprising the body-in-white

Hemmed/clinched joint = tight bend over similar thickness around doors or other closures

OEM = original equipment manufacturer

Other abbreviations, acronyms and technical references referred to in the text are presented in tabular form in Table 1.1.

Some familiarity with basic materials technology is assumed but for those readers requiring an introduction to subjects such as dislocation theory (referred to in Chapter 2), or broader coverage of mechanical properties (Chapter 5) reference is made to *Materials for Engineering* by W. Bolton,[6] an easily digested text, covering both metallic and non-metallic materials.

Table 1.1 Common abbreviations and acronyms used in the text

Technology	Terminology	Abbreviation
Steel condition	Annealed last	AL
	Skin passed or temper rolled	SP or TR
	Continuously cast	Concast
Sheet surface finish	Shot blasted	SB
	Electron beam textured	EBT
	Electro discharge textured	EDT
	Laser textured	Lasertex
	Mill finish (aluminium)	MF
	External surface standard, suitable for paint finish	Full finish or FF Class 'A' (plastics)
	Stretcher-strain free	SSF (aluminium)
Common steel grades	High strength steels	HSS
	Ultra high strength steel	UHSS
	Dual-phase	D-P
	Bake hardening	BH
	Transformation induced multi-phase steels	TRIP
	Interstitial free	IF
Coated steel types	Electrogalvanized	EZ or EG
	Hot-dip galvanized	HDG
	Iron–zinc alloy coated or Galvanneal	IZ
	Duplex galvanized, e.g. primer coated electrogalvanized steel	Bonazinc or Durasteel
Polymer abbreviations	Polyethylene	Polythene
	Acrylonitrile-butadiene-styrene terpolymer	ABS

Table 1.1 *(Contd)*

Technology	Terminology	Abbreviation
	Phenol formadehyde	Bakelite
	Polyvinyl chloride	PVC
	Polyamides	Nylon
	SP Resin Infusion Technology	SPRINT
Mechanical properties	Yield stress	YS
	Ultimate tensile stress	UTS
	Elongation	% El or elongn.
	Drawability index	'r' value
	Work-hardening index	'n' value
Major materials and related development programmes	Energy Conservation Vehicle	ECV
	Aluminium Structured Vehicle Technology	ASVT
	Audi Space Frame	ASF
	UtraLight Steel AutoBody	ULSAB
Recycling initiatives/ terminology	Automotive COnsortium on Recycling and Disposal	ACORD
	End of Life (of) Vehicles	ELV
	Authorized Treatment Facilities	ATF
	Certificate of Destruction	CoD
	Auto Shredder Residue or 'fluff'	ASR

References

1. Dieffenbach, J.R., 'Challenging Today's Stamped Steel Unibody: Assessing Prospects for Steel, Aluminium and Polymer Composites', IBEC '97 Proceedings Stuttgart, Germany, 30 Sept. – 2 Oct. 1997, pp. 113–118.
2. Ludke, B., 'Functional Design of a Lightweight Body-in-White Taking the New BMW Generation as an Example', *Stahl und Eisen*, 119, 12 May 1999, pp. 123–128.
3. Davies, G.M. and Easterlow, R., 'Automotive Design and Materials Selection', *Metals and Materials*, January 1985, pp. 20–25.
4. Diamant, R.M.E., *Rust and Rot and What You Can Do About Them*, Angus and Robertson, London, 1972.
5. Davies, G.M. and Goodyer, B., 'Aluminium in Automotive Applications', *Metals and Materials*, February 1993, pp. 86–91.
6. Bolton, W., *Materials for Engineering*, Newnes, Elsevier, Oxford, second edition, 2000.

2 Design and material utilization

Objective: To briefly review the historical development of the automotive body structure before considering how materials have helped to realize engineering and associated objectives (manufacturing objectives are considered in Chapter 5). Specific examples are selected to illustrate the changes that have taken place in the selection of steel grades, the emergence of aluminium and the increasing trend to hybrid material combinations.

Content: A brief outline is given on the evolution of various design concepts and materials utilization – reference is made to specific examples, where relevant changes have been made – the criteria typically used in optimizing design using FEM are defined – the utilization of coated and high strength steel grade are introduced – alternative body architecture is considered – use of aluminium and other lightweight materials are referenced together with the use of hybrid structures.

2.1 Introduction

While highlighting significant developments in the design architecture of the automotive body structure, the emphasis of this chapter is more concerned with the selection and use of materials, and how engineers have utilized relevant properties to satisfy their selection criteria over a timescale which now covers at least 100 years. Early materials selection was fairly limited and chiefly dictated by cost as the demands of mass production grew. Later availability became an issue, as two world wars had a draining effect on resources. But perhaps the period of greatest interest has spanned the last 30 years. During this time engineers have had to respond to a series of different outside influences (as will become apparent from Chapter 8) among them legislation on safety, emissions control and recycling. These have resulted in some conflicting results, for while weight reduction and use of lower density materials have extended the choice the increased variety and particularly the use of some plastics have caused additional headaches for the dismantling industry, as will be seen in Chapter 8.

After a consideration of milestones in autobody design through the last 100 years, the rationale of the modern designer is introduced with an example (from BMW) of how this is influenced by material choice. Examples are then given of alternative approaches to design using lightweight alternatives to steel, the traditional choice. Reference is made here to broader initiatives such as the major aluminium programmes (ECV, ASVT) and steel (ULSAB), which have demonstrated the feasibility of the newer technology. Together with relevant concept and competition car developments these are considered of sufficient importance to warrant a separate chapter (Chapter 4, p. 99). The spin-off from these background 'enabling' projects has been adopted in many parallel model programmes in recent years and this is self-evident in some of the instances given in this chapter. The same applies to many international initiatives (e.g. BRITE Light Weight Vehicle Programme BE – 5652[1]) plus supplier/user projects which have provided increased confidence in longer range concerns, and although the immediate pay-off from these programmes is not always immediately obvious it is critically important that funding for these general 'feeder' activities continues.

2.2 Historical perspective and evolving materials technology

The progess made in the development of engineering structures over the last century has been dealt with expertly elsewhere[2] and with regard to recent model programmes the significant use and benefits of FEM techniques (basic introduction summarized below) in shortening delivery times is emphatic.

However, as these become more complex, the more need there is for input detail such as material properties. As well as physical properties the need also exists for empirical data regarding material behaviour in diverse engineering situations and it is important that past designs and associated materials performance is analysed and 'rules' extracted for future design purposes and use in numerical form when required. In general terms the same developments have been evident on a worldwide scale, although size is a feature of American built vehicles, and the needs of mass production technology, reaching global proportions, have perhaps influenced the Japanese design philosophy (robot access, automation). Therefore although written from a UK perspective, with its foreign 'transplant' influences over the years the following content probably mirrors the worldwide trends and requirements for body materials in the future.

2.2.1 Body zones and terminology

First, it is necessary to clarify the terminology used to differentiate the various areas comprising the body. The body-in-white has already been introduced in Chapter 1 and this splits down into the main structure 'body-less-doors' and the 'bolt-on' or skin assemblies. These in turn break down into the inner panel, usually deep drawn to provide bulk shape and rigidity, plus the shallow skin panels which provide the outer contour of the body shape and require more aesthetic properties such as smooth blemish-free surface and scuff or dent resistance. The key elements of the main structure are the floor and main cage containing 'A', 'B/C' and 'D' posts or corner pillars and roof/cantrail surround, plus closed sections such as cross members, and front and rear longitudinal sections which provide essential impact resistance. The requirements of each zone are summarized in Table 2.1 together with recommendations for appropriate steels and possible alternatives:

2.2.2 Distinction between body-on-chassis and unitary architecture

Prior to the 1930s the body-on-chassis was the most popular vehicle configuration, the upper passenger containing compartment being mounted on a stout chassis which also carried the power-train unit plus other essential suspension, braking and steering gear. The body and chassis arrangement provided some versatility of model change and facility flexibility within the limited confines of earlier factories. Bodywork such as that used for the Morris Oxford in the early 1920s featured a wood, fabric and metal construction, the main change being to an all steel assembly in 1929 as the influence of the American Budd Company became obvious within Pressed Steel who supplied the body. The first significant aluminium body, the Pierce Arrow, also made its appearance in the early 1920s, but all steel construction had found favour in the USA because it was more suited to mass production, chiefly due to the ease of pressed panel production allied to the advantages of joining by spot welding. From an engineering point of view it also significantly increased torsional stiffness. A step change in design came with the integration of the chassis and body, claimed to have been introduced by Citroen in 1934 for its 11 CV model.[2] The difference in construction of the integrated or unitary construction compared with the chassis mounted body is illustrated by reference to the two modern day vehicles shown in Fig. 2.1.

Table 2.1   Requirements of different panels comprising the BIW structure

Zone/Assembly	Requirements	Materials choice		Possible* alternatives (material/form)
		Steels		
		Type	YS MPa	
Main structure:				
• Front/rear longit mbrs	Impact resistance	HSS	300	DP600, AP
• 'A' post inner/outer	Rigidity, strength	HSS	300	AP, HT
• Cantrail	,, ,,	HSS	260	AP, HT
• Main/rear floor	Moderate strength	HSS	180	AS
• Bodyside	Moderate strength, formability	HSS	180	AS, TWB
• Spare wheel well	Deep drawability	FS	140	SWS, SPA
• Wheelhouse, valance	Formability	FS	140	AS
'Bolt-on' assemblies				
Outer panels				
Door skins	'A' Class surface Dent resistance	FS	140	BH180, AS, SPA PLA-RRIM,
Bonnet	'A' Class surface Dent resistance	FS	140	BH180, AS, SPA PLA-SMC
Boot	'A' Class surface Dent resistance	FS	140	BH180, AS, SPA PLA-SMC
Roof	'A' Class surface Dent resistance	FS	140	BH180, HS
Inner panels				
Doors	Drawability	FS	140	TWB
Intrusion beams, rails	High impact strength	UHS	1200	AT, DP600+

* Code:
HSS = high strength steel; UHS = ultra HS steel; FS = forming steel; AS = aluminium sheet; AP = aluminium profile; SPA = superplastic Al; HT = hydroformed tube; HS = hydroformed sheet; BH180 = Bake hardened steel; Dpyyy = Dual phase steel; TWB = Tailor welded blank; PLA-xxx = Plastic-xxx type; RRIM = reinforced reaction injection moulded; SMC = sheet moulding compound; SWS = sandwich steel

2.2.3   Early materials and subsequent changes

Wood used in conjunction with fabric has been referred to already and was the construction of the bodywork of many cars in the 1920s before its replacement by steel. For outer panels this was of fairly thick gauge between 0.9 and 1.00 mm and much of it destined for the UK Midlands car plants was produced in the South Wales steelworks in ingot cast rimming or stabilized grades (Chapter 3). The rimming steels could be supplied in the 'annealed last' condition for deeper drawn internal parts but for surface critical panels a final skin pass was essential to optimize the paint finish. For complex and deeper drawn shapes the more expensive stabilized or aluminium killed material was used which conferred enhanced formability. Gradually a change took place – due to weight and cost reduction studies the average thickness of external panels reducing progressively to 0.8 mm in the 1950s/1960s and to the current level of 0.7 mm in use today for the production of the body of unitary construction shown in Fig. 2.1. Internal parts for structural members range from 0.7 to 2.0 mm, the scope for downgauging over the years being limited by stiffness constraints. Therefore although the thickness of strength related parts such as longitudinal members can be reduced by utilizing high strength grades on the basis of added impact resistance, as rigidity is a major design criterion, and the elastic

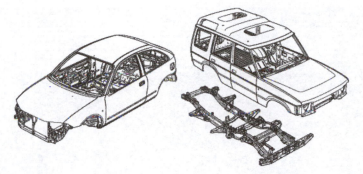

Fig. 2.1   Illustration of unitary and chassis body architecture[3]

modulus of steels is constant throughout the strength range, opportunities for substituting lighter gauges are limited. This situation can, however, be improved by use of adhesives or peripheral laser welded joints and examples of the use of these techniques are given later in this chapter and in Chapters 4 and 6.

Although not introduced until 1948 the Land Rover provides a good example of a modern vehicle with a chassis of two standard lengths serving a myriad of agricultural and military purposes. Although answering the rugged off-road requirements of the 4×4 vehicle virtually any type of body shape could be tailormade and constructed without the need for a dedicated higher volume facility. When steel was difficult to obtain in sheet or coil form in 1948 the underbody frames were produced by welding together strips of steel cast off remnants and aluminium was used for many body panels.

Together with the BMW 328 Roadster (1936–1940) and the Dyna Panhard (1954), Rover and Land Rover were among the first users of aluminium in Europe, the

Land Rover Discovery Tdi (3 door)

Fig. 2.2   Selection of Land Rover vehicles and body types

Land Rover Defender 90 hard top Tdi

Land Rover Defender 110 county station wagon Tdi

Fig. 2.2 *(Contd)*

ubiquitous Defender models using the 3xxx series alloys for flatter panels with the Al–Mg 5xxx series being used in other applications, a wealth of experience being gained in pressing, assembly and paint pre-treatment and finishing. Although the chassis was cumbersome it was – and still is – ideal for mounting the extensive range of Land Rover Defender body variants. Until this day the hot rolled grades of steel are used (typically HR 4) but it is easy to see why efforts are being made to downscale these relatively massive ladder frames with consideration being given to using newer material in thinner gauges, e.g. high strength steels up to 300 N/mm^2 (TRIP steels up to 590 N/mm^2 are now being used for 80 chassis parts on the Mitsubishi Paquera). Design modifications must be made to accommodate the thinner gauges and consideration has already been given to alternative material forms such as hydroformed sections (described later), as referenced by the ULSAB process, which could be used to bolster stiffness and crashworthiness. Although better suited to more conventional car body design, the incorporation of tailored blanks again offers an alternative approach giving the engineer strengthening exactly where required and a further opportunity for parts consolidation/reduced weight. This enduring type of rugged and versatile design has persisted as it answers the diverse needs of military purchasers but it is not surprising that as fleet average economy targets are considered more critically the monocoque is now becoming more stringent for the more volume-oriented 4 × 4 vehicles – as featured by the Land Rover Freelander. Durability is satisfied by the use of hot-dip or iron–zinc alloy coating as steel substrates replace the use of expensive aluminium for outer panels (see Chapter 7) and the model features another material innovation in the selection of polymer front wings.

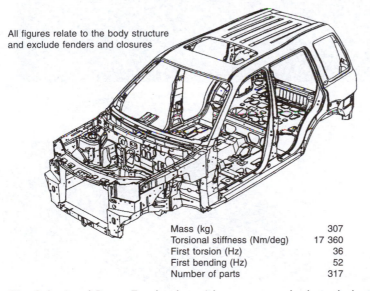

All figures relate to the body structure
and exclude fenders and closures

Mass (kg)	307
Torsional stiffness (Nm/deg)	17 360
First torsion (Hz)	36
First bending (Hz)	52
Number of parts	317

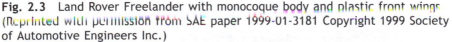

Fig. 2.3 Land Rover Freelander with monocoque body and plastic front wings (Reprinted with permission from SAE paper 1999-01-3181 Copyright 1999 Society of Automotive Engineers Inc.)

Before leaving body-on-chassis design it should be mentioned that other types of chassis include the steel backbone type used by Lotus and the designs featuring triangular sectional arrays as shown in Fig. 2.4. These were steel square or tubular sections, and later Lotus adopted another chassis configuration termed the 'punt', also shown.

Fig. 2.4 Selection of alternative chassis designs[2]

Fig. 2.4 *(Contd)*

The Lotus Elise featured the punt, which has also been termed a spaceframe concept and as this is more of a transitionary structure this will be described in greater detail later on together with similar aluminium internally structured bodies.

It has been debated as to what exactly constitutes a chassis-less design as various forms can incorporate some features of the original underframe, e.g. subframes and longitudinal/sidemember sections. Engineers such as Garrett[4] claim that the ideal form of chassis-less construction emerged in the 1940s with the launching of the Austin A30 as shown in Fig. 2.5.

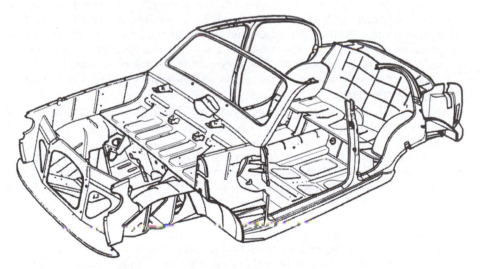

Fig. 2.5   Base structure of the Austin A30

They would argue that using their aircraft design principles they were able to incorporate all the essential load bearing requirements into a relatively lightweight body without even building in partial box sections that were featured in 'integral' or 'unitary' designs, with elements of the chassis incorporated in the underbody. However, even with such box sections and subframes the easily spot welded and finished bodies provided a significant advance in bodyweight reduction while meeting most engineering and manufacturing criteria.

The unitary design (monocoque is referred to in the industry but some say this should be reserved for competition type bodies of tube configuration) is by far the most popular type of body and using the powerful FEM analytical programs that exist today (see below) the design can be optimized at the design stage to maximize the use of properties and thereby reduce the number of prototypes, rework and development time. The more numerical data that can be gathered at this stage related to materials behaviour the more efficient modelling will be, and this applies to other simulation processes besides those predicting dynamic and static behaviour such as impact and torsional stiffness (demonstrated later). Forming is the obvious example but the complexity of accurately predicting thin shell behaviour during pressing brings in other variables as well as mechanical properties including friction, lubrication and topography.

2.3 Finite element analysis

For those readers requiring some basic understanding of FEA which is now a standard feature of computer-aided design (CAD) procedures used by body designers the following extract is presented from the publication 'Lightweight Electric/Hybrid Design' by Hodkinson and Fenton.[5]

This computerized structural analysis technique has become the key link between structural design and computer-aided drafting. However, because the small size of the elements usually prevents an overall view, and the automation of the analysis tend to mask the significance of the major structural scantlings, there is a temptation to by-pass the initial stages in structural design and perform the structural analysis on a structure which has been conceived purely as an envelope for the electromechanical systems, storage medium, passengers and cargo, rather than an optimized load-bearing structure. However, as well as fine-mesh analysis which gives an accurate stress and deflection prediction, course-mesh analysis can give a degree of structural feel useful in the later stages of conceptual design, as well as being a vital tool at the immediate pre-production stage.

One of the longest standing and largest FEA software houses is PAFEC who have recommended a logical approach to the analysis of structures, Fig. 2.6. This is seen in the example of a constant-sectioned towing hook shown at (a). As the loading acts in the plane of the section the elements chosen can be plane. Choosing the optimum mesh density (size and distribution) of elements is a skill which is gradually learned with experience. Five meshes are chosen at (b) to show how different levels of accuracy can be obtained.

The next step is to calculate several values at various key points – using basic bending theory as a check. In this example nearly all the meshes give good displacement match with simple theory but the stress line-up is another story as shown at (c). The lesson is: where stresses vary rapidly in a region, more densely concentrated smaller elements are required; over-refinement could of course, strain computer resources.

Each element is connected to its neighbour at a number of discrete points, or nodes, rather than continuously joined along the boundaries. The method involves setting up relationships for nodal forces and displacements involving a finite number of

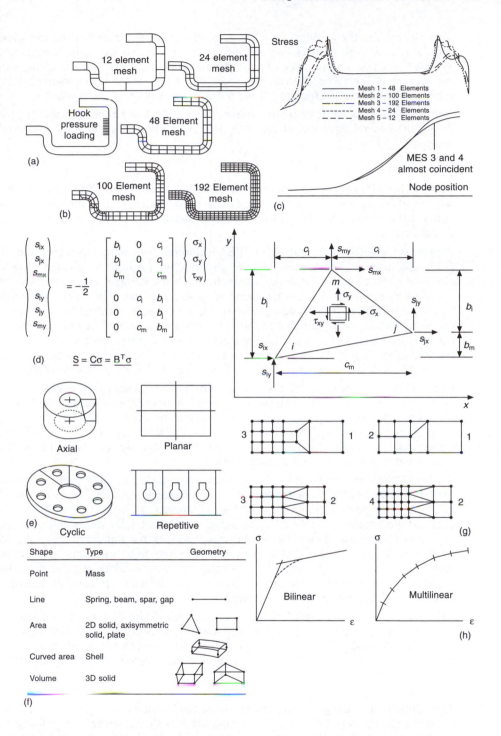

Fig. 2.6 Development of FEA: (a) towing hook as structural example;
(b) various mesh densities; (c) FEA vs elasticity theory; (d) node equations in matrix
form; (e) types of symmetry; (f) element shapes; (g) varying mesh densities;
(h) stress–strain curve representation

simultaneous linear equations. Simplest plane elements are rectangles and triangles, and the relationships must ensure continuity of strain across the nodal boundaries. The view at (d) shows a force system for the nodes of a triangular element along with the dimensions for the nodes in the one plane. The figure shows how a matrix can be used to represent the coefficients of the terms of the simultaneous equations.

Another matrix can be made up to represent the stiffness of all the elements $[K]$ for use in the general equation of the so-called 'displacement method' of structural analysis:

$$[R] = [K] \cdot [r]$$

where $[R]$ and $[r]$ are matrices of external nodal forces and nodal displacements; the solution of this equation for the deflection of the overall structure involves the inversion of the stiffness matrix to obtain $[K]^{-1}$. Computer manipulation is ideal for this sort of calculation.

As well as for loads and displacements, FEA techniques, of course, cover temperature fields and many other variables and the structure, or medium, is divided up into elements connected at their nodes between which the element characteristics are described by equations. The discretization of the structure into elements is made such that the distribution of the field variable is adequately approximated by the chosen element breakdown. Equations for each element are assembled in matrix form to describe the behaviour of the whole system. Computer programs are available for both the generation of the meshes and the solution of the matrix equations, such that use of the method is now much simpler than it was during its formative years.

Economies can be made in the discretization by taking advantage of any symmetry in the structure to restrict the analysis to only one-half or even one-quarter – depending on degree. As well as planar symmetry, that due to axial, cyclic and repetitive configuration, seen at (e), should be considered. The latter can occur in a bus body, for example, where the structure is composed of identical bays corresponding to the side windows and corresponding ring frame.

Element shapes are tabulated in (f) – straight-sided plane elements being preferred for the economy of analysis in thin-wall structures. Element behaviour can be described in terms of 'membrane' (only in-plane loads represented), in bending only or as a combination entitled 'plate/shell'. The stage of element selection is the time for exploiting an understanding of basic structural principles; parts of the structure should be examined to see whether they would typically behave as a truss frame, beam or in plate bending, for example. Avoid the temptation to over-model a particular example, however, because number and size of elements are inversely related, as accuracy increases with increased number of elements.

Different sized elements should be used in a model – with high mesh densities in regions where a rapid change in the field variable is expected. Different ways of varying mesh density are shown at (g), in the case of square elements. All nodes must be interconnected and therefore the fifth option shown would be incorrect because of the discontinuities.

As element distortion increases under load, so the likelihood of errors increases, depending on the change in magnitude of the field variable in a particular region. Elements should thus be as regular as possible – with triangular ones tending to equilateral and rectangular ones tending to square. Some FEA packages will perform distortion checks by measuring the skewness of the elements when distorted under load. In structural loading beyond the elastic limit of the constituent material an idealized stress/strain curve must be supplied to the FEA program – usually involving a multilinear representation, (h).

When the structural displacements become so large that the stiffness matrix is no longer representational then a 'large-displacement' analysis is required. Programs can include the option of defining 'follower' nodal loads whereby these are automatically reorientated during the analysis to maintain their relative position. The program can also recalculate the stiffness matrices of the elements after adjusting the nodal coordinates with the calculated displacements. Instability and dynamic behaviour can also be simulated with the more complex programs.

The principal steps in the FEA process are: (i) idealization of the structure (discretization); (ii) evaluation of stiffness matrices for element groups; (iii) assembly of these matrices into a supermatrix; (iv) application of constraints and loads; (v) solving equations for nodal displacements; and (vi) finding member loading. For vehicle body design, programs are available which automate these steps, the input of the design engineer being, in programming, the analysis with respect to a new model introduction. The first stage is usually the obtaining of static and dynamic stiffness of the shell, followed by crash performance based on the first estimate of body member configurations. From then on it is normally a question of structural refinement and optimization based on load inputs generated in earlier model durability cycle testing. These will be conducted on relatively course mesh FEA models and allow section properties of pillars and rails to be optimized and panel thicknesses to be established.

In the next stage, projected torsional and bending stiffnesses are input as well as the dynamic frequencies in these modes. More sophisticated programs will generate new section and panel properties to meet these criteria. The inertias of mechanical running units, seating and trim can also be programmed in and the resulting model examined under special load cases such as pot-hole road obstacles. As structural data is refined and updated, a fine-mesh FEA simulation is prepared which takes in such detail as joint design and spot-weld configuration. With this model a so-called sensitivity analysis can be carried out to gauge the effect of each panel and rail on the overall behaviour of the structural shell.

Joint stiffness is a key factor in vehicle body analysis and modelling them normally involves modifying the local properties of the main beam elements of a structural shell. Because joints are line connections between panels, spot-welded together, they are difficult to represent by local FEA models. Combined FEA and EMA (experimental modal analysis) techniques have thus been proposed to 'update' shell models relating to joint configurations. Vibrating mode shapes in theory and practice can thus be compared. Measurement plots on physical models excited by vibrators are made to correspond with the node points of the FEA model and automatic techniques in the computer program can be used to update the key parameters for obtaining a convergency of mode shape and natural frequency.

An example car body FEA at Ford was described at one of the recent Boditek conferences, Fig. 2.7, outlining the steps in production of the FEA model at (a). An extension of the PDGS computer package used in body engineering by the company – called FAST (Finite-Element Analysis System) – can use the geometry of the design concept existing on the computer system for fixing of nodal points and definition of elements. It can check the occurrence of such errors as duplicated nodes or missing elements and even when element corners are numbered in the wrong order. The program also checks for misshapen elements and generally and substantially compresses the time to create the FEA model.

The researchers considered that upwards of 20 000 nodes are required to predict the overall behaviour of the body-in-white. After the first FEA was carried out, the deflections and stresses derived were fed back to PDGS-FAST for post-processing.

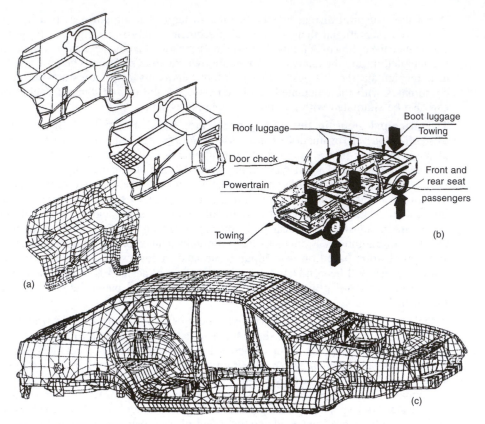

Fig. 2.7 FEA of Ford car: (a) steps in producing FEA model; (b) load inputs; (c) global model for body-in-white (BIW)

This allowed the mode of deformation to be viewed from any angle – with adjustable magnification of the deflections – and the facility to switch rapidly between stressed and unstressed states. This was useful in studying how best to reinforce part of a structure which deforms in a complex fashion. Average stress values for each element can also be displayed numerically or by graduated shades of colour. Load inputs were as shown at (b) and the FE model for the BIW at (c).

2.4 One manufacturer's approach to current design

It is now timely to consider the more contemporary approach to design and reference is made to the approach that BMW have adopted to utilizing materials to optimize structural performance while at the same time satisfying prevailing safety, performance and environmental requirements. Their progress is illustrated by extracts taken from information presented recently by Bruno Ludke, BMW Body Design specialist, at recent international automotive conferences.

2.4.1 Product requirements

In terms of lightweight bodyshell functional design Ludke[6] has identified four areas for critical consideration:

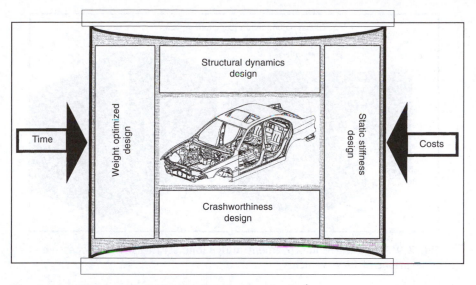

Fig. 2.8 Critical areas for body design (after Ludke[6])

- Structural dynamics
- Static stiffness
- Crashworthiness
- Weight optimization

2.4.2 Structural dynamics

Improvements in performance including significant weight savings in the steel body achieved over recent model generations are described in the following sections and these are attributed to the effective application of FEM analysis and the interrelationship with material properties.

Structural dynamics is described[6] as the achievement of the desired level of comfort in terms of noise, vibration and harshness (NVH) for which the yardstick is taken as behaviour at idling speed – normally between 600 and 700 rpm. To ensure 'vibration-free' operation, the frequencies for the first bending and torsional natural modes of the complete vehicle must lie within a limited frequency range. The upper limit of this range is represented by the third engine order of the 6-cylinder engine and the lower limit by the second engine order of the 4-cylinder engines thus constituting an 'idle frequency window'. To attain the target frequencies of 26/29 Hz for the vehicle as a whole, the corresponding natural modes of the bodyshell must be twice as high and no local modes must occur below these frequencies, e.g. at the front or rear of the vehicle. The improvements achieved with the outstandingly popular 3, 5 and 7 series BMW models are illustrated schematically in Fig. 2.9 and the success in this area is attributed primarily to the application of FEM analysis and experimental modal analysis technique applied at the early stages of body shell development.

2.4.3 Design for static stiffness

Static design entails the optimization of torsional stiffness and strength under quasistatic loading conditions and good static stiffness values are fundamental requirements for target dynamic characteristics previously described. The variation in torsional stiffness

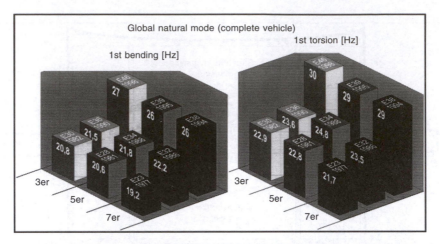

Fig. 2.9 Progressive improvement in dynamic stiffness with successive generations of BMW vehicles

with vehicle kerb weight (K_w) has been developed as shown in Fig. 2.10 for BMW models, the target C_t value being $15 \times K_w$. To avoid excessive loading of the windscreen and stone chipping damage resulting from excessive surface stresses, the inherent stiffness without glass must reach 66 per cent of the final stiffness. Specific design improvements were made in the latest models to key joints and structural members to increase torsional stiffness from 20 000 Nm/° to 28 500 Nm/°. Again the progression through successive BMW models is shown in Fig. 2.10 with a doubling of previous values.

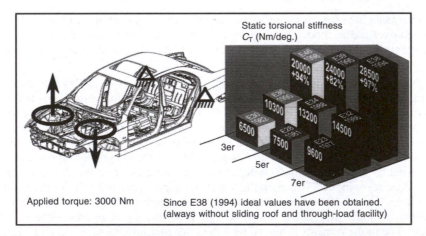

Fig. 2.10 Progressive improvement in torsional stiffness shown for successive BMW model generations[6]

2.4.4 Crashworthiness

All vehicle manufacturers are placing continued emphasis on occupant passive safety and here FEM simulation is of special importance, avoiding the need for expensive vehicle compliance tests during development. In the case of more recent models referred to above the stiffness and dynamic improvements form an excellent basis for crash optimization, and as requirements are aligned to 40 mph impacts the absorbed energy per structural unit (vehicle side) has risen by 80 per cent in comparison with predecessors. The shift in design requirements over the last 25 years is illustrated in Figs 2.11 and 2.12 together with the configurations used for modelling 40 mph offset crash and side impact simulations.

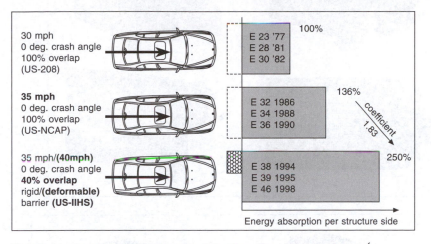

Fig. 2.11 Increasingly stringent objectives for crashworthiness[6]

2.4.5 Weight efficiency

Although a basic design requirement previously, the drive for lower weight vehicles, in the knowledge that 10 per cent reduction in vehicle mass leads to fuel savings of up to 6–7 per cent has intensified over the last 20 years. In the 1970s and 1980s the initial momentum swung towards aluminium as the industry attempted to confirm the fuel economy figures using the most radical materials solution available at the time. The ECV and ASV programmes (referred to in Chapter 4) together with the aluminium structured A8, A2, NSX and Z8 programmes described elsewhere in this chapter, have helped confirm more efficient consumption figures but have also underlined the substantial changes in the supply and process chain manufacturing facilities required together with peripheral costs such as higher repair and subsequently insurance costs. It is these factors which may account for the slow emergence of aluminium as a significant body material despite more positive forecasts and the fact that most major organizations have now gained the technological and design experience (with low volume derivatives) to enter full-scale production.

The feeling also existed in the early 1990s that a lot more potential for weight reduction still existed with steel albeit in slightly different guises which would yield useful weight savings, and perhaps if mixed with lighter materials such as aluminium or plastic skins would allow most future objectives to be achieved. The more flexible facilities and experience gained over recent years could be adapted to accept newer high strength materials and different configurations such as TWBs and hydroformed tube sections. It was with this knowledge that the steel industries response in the 1990s has been a design study undertaken on behalf of 32 steel producers by Porsche Engineering Services (the ULSAB programme described in Chapter 4).

The significance of bodyweight on fuel consumption, acceleration and emissions control has been outlined already, but to put these in perspective with other relevant factors some of these parameters are illustrated in Fig. 2.13.

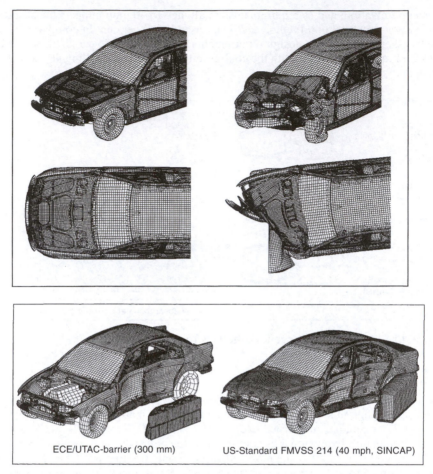

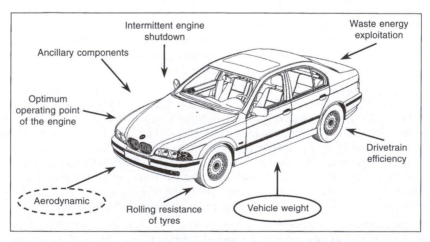

ECE/UTAC-barrier (300 mm)   US-Standard FMVSS 214 (40 mph, SINCAP)

Fig. 2.12 Offset barrier (top) and side impact simulations (below)[6]

Fig. 2.13 Factors contributing to improved fuel economy[6]

Despite the improved functionality already described, the optimized unitary design of the body shell resulted in it making up a significantly reduced proportion of the kerbweight as shown in Fig. 2.14.

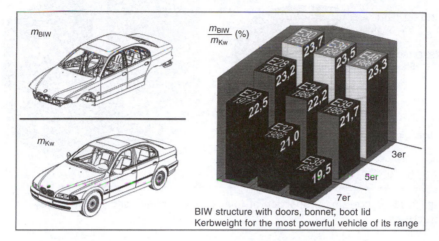

Fig. 2.14   Relationship of body-in-white weight to kerbweight[6]

A factor 'L' has been used by BMW to summarize the weight reduction improvement effected by design, which relates to structural performance and vehicle size and is shown in Fig. 2.15 together with the progressive achievement over the years and although relatively empirical does provide a measure of design optimization.

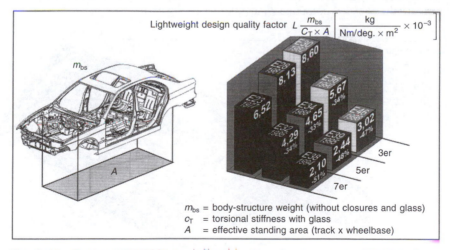

Fig. 2.15   Design efficiency as defined by functional optimization and size, over three generations of vehicles[6]

More specific materials related data was further presented by Ludke[7] who referred to the changes in HSS utilization which had accompanied the functional improvements in the various BMW models referred to above.[6,7] As shown in Fig. 2.16 the proportion of high strength steel was increased from 4.5 per cent for the 5-Series model to 50 per cent. This utilized the range of bake-hardening steels H180B to H300B together with

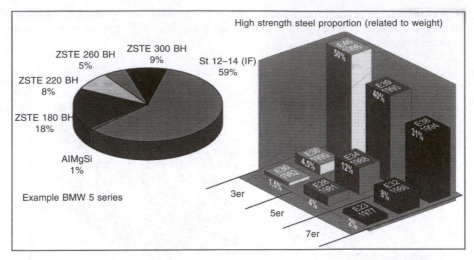

Fig. 2.16 Increasing proportion of high strength steel used in BMW body structures[7]

isotropic and IF HSS grades. The increased utilization of HSS grades is typical of strengths now being incorporated in current designs by European body engineers using the full range of rephosphorized, IF HS, HSLA and bake hardening grades which are included in Euronorm 10292. This covers hot-dip galvanized grades, again reflecting the improvements and change in durability required by today's structures, but an uncoated parallel standard is in preparation.

The principal parameters (referred to in Table 3.2, p. 63) used in the analysis of the structural and panel components are shown in Fig. 2.17.

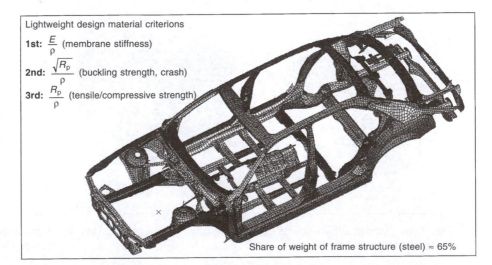

Fig. 2.17 Criteria used in the analysis of structural components by BMW[6]

Similar analyses were carried out on bolt-on assemblies but totally different criteria apply as will become evident in the following section.

Evaluation of individual requirements of each part are made at an early simulation

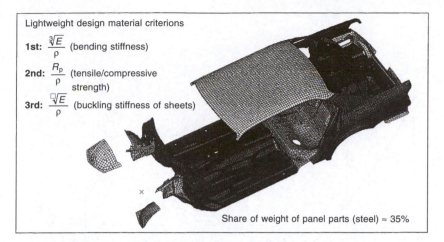

Lightweight design material criterions

1st: $\dfrac{\sqrt[3]{E}}{\rho}$ (bending stiffness)

2nd: $\dfrac{R_p}{\rho}$ (tensile/compressive strength)

3rd: $\dfrac{\sqrt{E}}{\rho}$ (buckling stiffness of sheets)

Share of weight of panel parts (steel) ≈ 35%

Fig. 2.18 Criteria used in the analysis of panel parts by BMW

stage. One method of enhancing stiffness is to use linear laser welding or apply a structural adhesive to inner flange or seam surfaces and this is a key development for the future. It is generally recognized that a compatible pretreatment such as those used in aircraft construction is required to ensure that any degradation of the bond does not occur in service and it is equally important to maintain coating integrity through forming and assembly to ensure a firm foundation for the adhesive system.

As described previously the influence of material properties on impact and collapse characteristics is becoming more evident with the development of dual-phase and TRIP steels which, due to unique work hardening and ductility combinations (increased area under the stress/strain curve), offer increased energy absorption. (Instances where these steels are being exploited are given in Chapter 8, under section 8.5.) Again enhancement of these properties should be possible in combination with adhesive application leading to a high strength steel utilization of 80–90 per cent.

2.5 Panel dent resistance and stiffness testing

Optimized designs of outer body panels must also meet several other performance criteria including stiffness, oil canning or critical buckling load and dent resistance. Stiffness is a fundamental concern for the perceived quality of a body panel. Along with oil canning (the 'popping' of a panel when pressed) it determines how the panel 'feels' to a customer. Dent resistance is important to avoid panel damage in-plant and minimize dents and dings on external parts in-service. Poor panel quality in used cars will generally depress resale values and possibly influence the decision to purchase a particular brand.

From a practical point of view, dents can be caused in a number of ways and on the full range of external body panels. Considering doors as an example, denting can occur from stone impacts (dynamic denting) or, to the frustration of the vehicle owner, from the careless opening of an adjacently parked vehicle door. Denting can occur where the door surface is smooth and may not have sufficient curvature to resist 'door slamming' (quasi-static denting) or along prominent feature lines where 'creasing' can occur.

Panel dent resistance and stiffness has been the subject of considerable research. Despite this there is no industry-wide, generally adopted method of testing. For

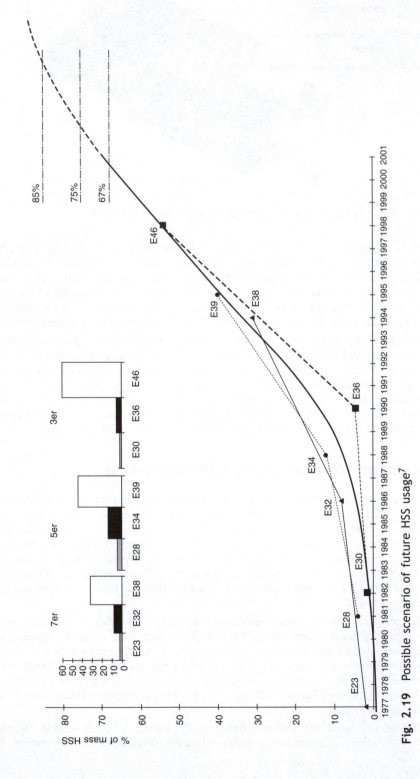

Fig. 2.19 Possible scenario of future HSS usage[7]

quasi-static dent testing, a wide range of purpose built dent resistance/stiffness test equipment configurations have been employed within the automotive industry. In addition, the configuration of a tensile testing machine for compression testing and similar modified equipment has allowed suitable data to be obtained. Whichever system is used, the principle of force application resulting in deflection and ultimately plastic deformation of the panel remains the same. Variables can include method of load application (hydraulic or stepper motor), speed of load application, indenture shape and size and panel assembly conditions. Some reported methods of testing are based on repeated application and removal of force at increasing levels. Others involve the continued application of a steadily increasing force until denting occurs. In some cases, stiffness is assessed using the same basic test equipment but with a much larger radiussed loading head to prevent localized deformation. Force and displacement measurements are generally incorporated into a data acquisition system.

An illustration of a typical output from such a test would be as represented in Fig. 2.20. Initial stiffness is given by the slope of the curve in the first region, until the buckling load is reached. After 'oil canning', the panel continues to deflect elasticity, before the onset of plastic deformation in the material. When the load is reversed, the permanent deformation of the panel is indicated since the lower portion of the curve does not return to zero.

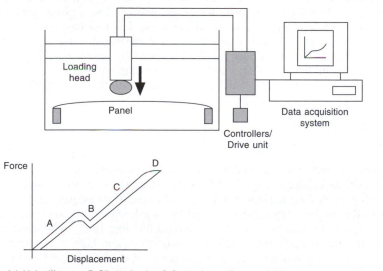

A Initial stiffness   B Oil can load   C Secondary stiffness   D Plastic deformation

Fig. 2.20 Test rig for measurement of panel dent resistance and stiffness

Experimental testing of dynamic dent resistance has previously concentrated on drop weight rigs, using various indentor masses and drop heights to achieve a range of denting energies. Even higher energies can be achieved through the use of a compressed air operated ball bearing gun. The key issue to be considered is that test conditions (denting force, impact speed, etc.) must be genuinely representative of those conditions existing in field, i.e. if the energy input to cause a perceptible dent in-service is 10 J, then the dent testing procedure should reflect this.

Test results generated from the above techniques will typically be compared against performance standards set by an individual manufacturer. Standards widely known include those published by the American Iron and Steel Institute which defines a minimum dent resistance of 9.7 J and a stiffness that should exceed 45 N/mm.

Based on testing using the practical techniques outlined, empirical formulae predicting the force and energy required to initiate a dent have been presented in recent years. Typically:

$$W = (K.YS^2.t^4)/S$$

where W is the denting energy, K is a constant, YS is the material yield strength, t is the panel thickness and S is the panel stiffness. Panel stiffness depends upon the elastic modulus, the panel thickness, shape and geometry and boundary conditions. The ability of plastic panels to meet light denting is a definite advantage (as long as 40 years ago Henry Ford could be seen striking a Ford development vehicle with plastic panels, to demonstrate the ability of plastic/composites to resist denting. Nonetheless, high strength steels offer more dent resistance at the same thickness as mild steel, or the opportunity for weight saving and equivalent dent resistance at reduced panel thickness.

Given the many iterations of automotive body panel design that can take place, it is usually late in the product development process that the first production representative parts are available for dent and stiffness testing. With press tooling already produced, it is generally only initial material properties that can be changed or local reinforcements added to improve the stiffness/dent resistance. It is not surprising, therefore, that currently much attention is being focused on the use of analytical tools such as finite element analysis, for body panel performance predictions. Thus, given certain part geometry and dimensions, predictions of stiffness and dent resistance can be made. Based on material gauge and grade and in the case of metallic panels, strain levels in the material, optimization of the design can take place. Should the accuracy of such techniques be proven, the use of dent and stiffness testing equipment may in future be limited to selected verification of such performance predictions and quality control issues.

2.6 Fatigue

The behaviour of sheet materials under conditions of constantly fluctuating stress or strain is of critical importance to the life body structure, whether of high or low frequency. High cycle fatigue is more descriptive of conditions existing, say, in close proximity to the engine compartment, while low cycle conditions represent those induced by humps and bumps encountered in road running. Both are assessed very carefully in the initial engineering selection procedure, the high cycle behaviour being determined using Wohler S–N curves, often used as the input to CAD design programmes. Steels typically give a clearly defined fatigue limit below which components can be designed in relative safety but aluminium gives a steady stress reduction with time. As described in Chapter 4, a cautionary note in using cold work strengthening – low cycle fatigue can induce a progressive cyclic softening, which can counteract the strengthening developed by strain ageing as well as cold deformation.

The behaviour of a particular design is very difficult to predict due to the nature of materials characteristics combined with the complexity of design features comprising all body shapes, and which can result in stress concentrations. Therefore, despite extensive measurements and predictive programmes the only true way to determine the sensitivity of a structure to cyclic behaviour is rig testing. This can take the form of simple push–pull load application or extend to four poster simulated movements gathered under arduous track testing. Push–pull tests even of the simple tensile test type must be carried out carefully to avoid buckling effects which may limit the range of thicknesses on which these tests may be used. If investment can be made in the hydraulic facilities necessary for the full rig simulation, these are the only realistic means of detecting weaknesses prone to cyclic failure apart of course from the accelerated track tests over rough terrain.

Weaknesses can be identified by the application of stress lacquer techniques or similar, and modification carried out by localized strengthening. The effect of material properties is debatable as again body features are claimed to negate these, especially when considering spot welded joints. Many studies have shown that with high strength steels the notch effects associated with the weld geometry overpower any effect due to material strength.

A more lengthy description of the fatigue process and body design follows and is borrowed from *Lightweight Electric/Hybrid Vehicle Design*[5] and presents a concise summary of factors applicable to fatigue resistance, relevant to most body structures.

2.6.1 Designing against fatigue

Dynamic factors should also be built in for structural loading, to allow for travelling over rough roads. Combinations of inertia loads due to acceleration, braking, cornering and kerbing should also be considered. Considerable banks of road load data have been built up by testing organizations and written reports have been recorded by MIRA and others. As well as the normal loads which apply to two wheels riding a vertical obstacle, the case of the single wheel bump, which causes twist of the structure, must be considered. The torque applied to the structure is assumed to be $1.5 \times$ the static wheel load $\times$ half the track of the axle. Depending on the height of the bump, the individual static wheel load may itself vary up to the value of the total axle load.

As well as shock or impact loading, repetitive cyclic loading has to be considered in relation to the effective life of a structure. Fatigue failures, in contrast to those due to steady load, can of course occur at stresses much lower than the elastic limit of the structural materials, Fig. 2.21. Failure normally commences at a discontinuity or surface imperfection such as a crack which propagates under cyclic loading until it spreads across the section and leads to rupture. Even with ductile materials failure occurs without generally revealing plastic deformation. The view at (a) shows the terminology for describing stress level and the loading may be either complete cyclic reversal or fluctuation around a mean constant value. Fatigue life is defined as the number of cycles of stress the structure suffers up until failure. The plot of number of cycles is referred to as an *S–N* diagram, (b), and is available for different materials based on laboratory controlled endurance testing. Often they define an endurance range of limiting stress on a 10 million life cycle basis. A log–log scale is used to show the exponential relationship $S = C. Nx$ which usually exists, for C and x as constants, depending on the material and type of test, respectively. The graph shows a change in slope to zero at a given stress for ferrous materials – describing an absolute limit for an indefinitely large number of cycles. No such limit exists for non-ferrous metals and typically, for aluminium alloy, a 'fatigue limit' of 5×10^8 is defined. It has also become practice to obtain strain/life (c) and dynamic stress/strain (d) for materials under sinusoidal stroking in test machines. Total strain is derived from a combination of plastic and elastic strains and in design it is usual to use a stress/strain product from these curves rather than a handbook modulus figure. Stress concentration factors must also be used in design.

When designing with load histories collected from instrumented past vehicle designs of comparable specification, signal analysis using rainflow counting techniques is employed to identify number of occurrences in each load range. In service testing of axle beam loads it has been shown that cyclic loading has also occasional peaks, due

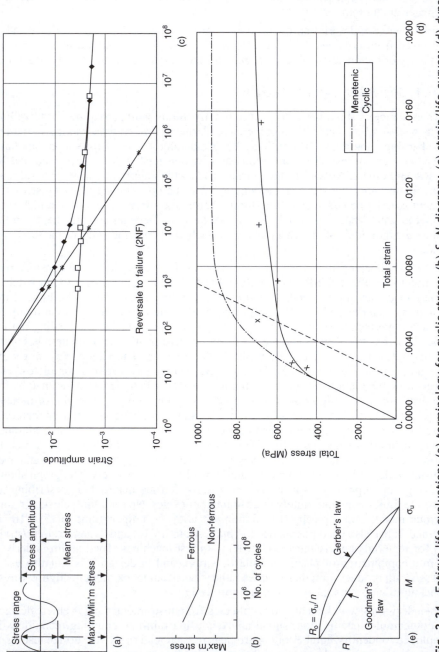

Fig. 2.21 Fatigue life evaluation: (a) terminology for cyclic stress; (b) S–N diagram; (c) strain/life curves; (d) dynamic stress/strain curves; (e) fatigue limit diagrams

to combined braking and kerbing, equivalent to four times the static wheel load. Predicted life based on specimen test data could be twice that obtained from service load data. Calculation of the damage contribution of the individual events counted in the rainflow analysis can be compared with conventional cyclic fatigue data to obtain the necessary factoring. In cases where complete load reversal does not take place and the load alternates between two stress values, a different (lower) limiting stress is valid. The largest stress amplitude which alternates about a given mean stress, which can be withstood 'infinitely', is called the fatigue limit. The greatest endurable stress amplitude can be determined from a fatigue limit diagram, (e), for any minimum or mean stress. Stress range R is the algebraic difference between the maximum and minimum values of the stress. Mean stress M is defined such that limiting stresses are $M +/-R/2$.

Fatigue limit in reverse bending is generally about 25% lower than in reversed tension and compression, due, it is said, to the stress gradient – and in reverse torsion it is about 0.55 times the tensile fatigue limit. Frequency of stress reversal also influences fatigue limit – becoming higher with increased frequency. An empirical formula due to Gerber can be used in the case of steels to estimate the maximum stress during each cycle at the fatigue limit as $R/2 + (\sigma_u 2 - nR\sigma_u)^{1/2}$ where σ_u is the ultimate tensile stress and n is a material constant = 1.5 for mild and 2.0 for high tensile steel. This formula can be used to show the maximum cyclic stress σ for mild steel increasing from one-third ultimate stress under reversed loading to 0.61 for repeated loading. A rearrangement and simplification of the formula by Goodman results in the linear relation $R = (\sigma_u/n) [1 - M/\sigma_u]$ where $M = \sigma - R/2$. The view in (e) also shows the relative curves in either a Goodman or Gerber diagram frequently used in fatigue analysis. If values of R and σ_u are found by fatigue tests then the fatigue limits under other conditions can be found from these diagrams.

Where a structural element is loaded for a series of cycles $n1$, $n2$ . . . at different stress levels, with corresponding fatigue life at each level $N1$, $N2$ . . . cycles, failure can be expected at $\Sigma n/N = 1$ according to Miner's law. Experiments have shown this factor to vary from 0.6 to 1.5 with higher values obtained for sequences of increasing loads.

2.7 Alternative body architecture

Before examples of more adventurous modern designs are presented, certain vehicles are now highlighted which illustrate further interesting steps in body and materials development. Having commenced with essentially steel bodies of unitary design, and it must be remembered that these still constitute the vast majority of volume cars produced, 'conventionally built' aluminium structures are considered, before moving to the spaceframe concept and finally the inevitable hybrid configurations. In this context hybrid means mixed material content and introduces the 'user friendly' advantages of polymers (low impact resistance and styling freedom) combined with the lightweight advantage of aluminium plus the practicality and safety connotations associated with steel. The latter has always been a strong argument of the anti-CAFE lobby in the USA who contend that the benefits in fuel economy accompanying hybrid lighterweight bodies are achieved at the expense of vehicle safety and claim to have accident statistics to prove this.

2.7.1 The unitary aluminium body

The development of the all aluminium body is now more associated with the A8 and A2 spaceframe type of vehicle (described later) which constitutes a different type of concept, and the need for a fundamentally different type of design may become more obvious if the production of an aluminium body is considered with conventional production technology. Although some use was made of aluminium prior to 1900 for the Durkopp developed sports car[8] and later reference was made to the Pierce Arrow body (1909) which incorporated rear end panel, roof, firewall and doors in cast aluminium, the Dyna Panhard was probably the first aluminium bodied car to be mass produced in Europe in any numbers. The Honda NSX sports car represents the most recent conventional body built within the context of modern manufacturing and proves that although some equipment modifications were necessary assembly was possible in moderate numbers.

2.7.1.1 The Honda NSX

Following a consideration of specific strength, specific rigidity and equivalent rigidity compared with sheet steel and SMC[9] the decision was made to manufacture the BIW in aluminium to reduce the weight by about 140 kg. The rigidity of a car such as the NSX is critical to maintain steering stability, and to help improve this the sills were produced as extrusions with variable side wall thickness. The comparison is shown in Fig. 2.22.

To satisfy different requirements for strength, formability, weldability and coating, detailed preparatory background studies showed that different alloys should be used for different panel applications and these are indicated in Fig. 2.23.

It was found that wrinkling and shape control were the main problems on forming, attributed to lower modulus which resulted in more springback (compared to steel) and also the lower 'r' value. Twice the overcrowning allowance was required than for steel in the forming of door outer panels. Together with proportionally lower forming limit curves, it was found that new disciplines in the form of die adjustments, crowning and lubrication were essential if the required shapes were to be mass produced.

Regarding welding, instantaneous welding currents of 20 000 to 50 000 amps were now necessary compared with 7000 to 12 000 amps for steel, and higher weld force values of 400 to 800 kgf compared with 200 to 300 kgf. A special hand welding gun was devised having a built-in small transformer to reduce current loss. Spot welds were augmented with short MIG welding runs. Prior to painting with a four-coat system, a change to a chromate–chromium pretreatment was found more suitable than the usual zinc phosphate formulation. Dacromet was found effective in protecting small steel parts, e.g. bolts, from bi-metallic corrosion.

2.7.2 The pressed spaceframe (or base unit) concept – steel

The use of aluminium skin panels was extended to the Rover car range, the P4 (1954–1964 doors, bonnet and boot lid), and P6 (1964–1976 bonnet and boot lid) although the SD1 body (Rover 3.5 1976–1984) reverted to steel. It was interesting to note that the P6 (Rover 2000) bonnet was changed to the 2117 Al–Cu grade to improve formability and obviate any signs of 'stretcher-strain' markings. However, the most significant design feature associated with the P6 was the appearance of the steel base unit in the 1964 P6 which featured a central frame to which pressed outer assemblies were rigidly and consistently bolted using drilled and tapped forged bosses.

This 'base unit' was then clad with steel fenders (wings) and doors while the bonnet and trunk lid (boot lid) were in aluminium. The advantage of this type of design is that in theory the cladding and external shape can be changed relatively frequently without changing the substructure, and repair simplified. This concept allowed clad panels in aluminium as an option and the same idea was adopted on the American

Fig. 2.22   Sill sections produced from pressed parts compared with extruded sections[9]

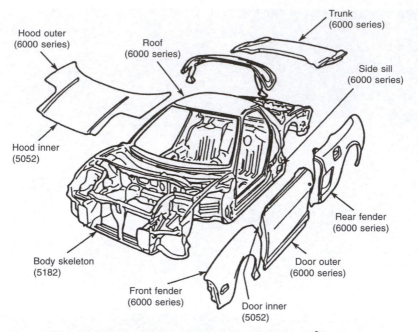

Fig. 2.23 Aluminium alloys used for NSX body panels[9]

Fig. 2.24 Rover P6 spaceframe

Pontiac Fiero, where a steel substructure was clad in polymer skin panels again using adjustable box type attachment points which could accommodate any differences in expansion between the two materials. The GM Saturn, shown in Fig. 2.25, has used the same type of pressed steel spaceframe (lower parts galvanized) for structural integrity and strength while clad in thermoplastic skin panels (doors, fenders, quarter panels and fascias) to enhance corrosion resistance and reduce damage from low speed impacts. The roof, bonnet and boot lid are retained in steel and the skin assemblies are painted in complete sets on support bucks in simulated on-car positions. The design technology was carried forward and developed with the Renault Espace which featured a steel substructure (in reality unibody structures with non-structural plastic cladding panels), but the distinction here was that the steel substructure was fully hot-dip galvanized prior to cladding with a polymer exterior, the penetration of zinc into crevices adding to the torsional stiffness of the main frame. This is theoretically

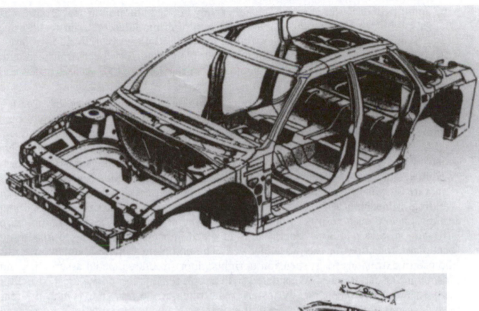

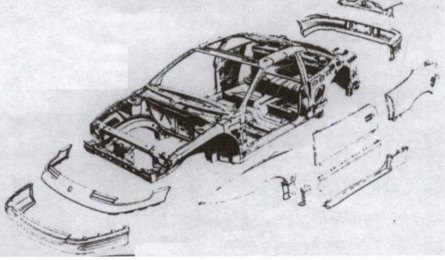

Fig. 2.25   Saturn spaceframe showing polymer panels on a four-door sedan[10]

the most effective type of corrosion protection allowing the encapsulation and full coverage of spot welds and cut edges, although thickness control can be variable and result in a weight penalty.

2.7.3 Examples of pressed aluminium spaceframes and associated designs

Again referring to examples of design innovation within the Rover vehicle range, the experimental ECV3 vehicle (see Chapter 4) had demonstrated that the base unit concept could be extended further to provide an even lighter structure using aluminium pressed parts. The torsional stiffness in that case was improved by the use of adhesive in a weldbonding mode employing a specially developed pretreatment and prelubrication technology. The manufacturing feasibility of this approach was proven by the production of a small fleet of Rover Metros. The adhesively bonded aluminium spaceframe was clad in plastic, the horizontal panels being in a high modulus material to improve flexure and sagging effects, with the vertical panels in RRIM polyurethane to improve low speed impact and denting. Similar technology has now been transferred to production vehicles via ASV designs applied to the Jaguar 220, the XJ series and Lotus Elise.

Introduced in September 1995, the Lotus Elise featured a further type of structure termed 'the punt'. This followed joint design technology developed by Lotus Engineering and Hydro Aluminium Automotive Structures of Denmark and as shown in Fig. 2.26 features aluminium extrusions joined by a combination of adhesive bonding and mechanical fasteners. At 68 kg the spaceframe achieved a 50 per cent weight reduction compared with an equivalent steel construction and with bonded structures it was found that thinner sections could be used, and compared with spot welding or mechanical fastening no local stresses are produced. Excellent torsional rigidity at low mass results in good driving force and agility[11] and the aluminium structure absorbs additional energy in high speed impacts contributing to maximum occupant protection for the passenger cell. The complete vehicle is noteworthy for the use of extrusions for suspension uprights, door structures, pedal assemblies and dashboard fascia. Repair is considered with a replacement composite crash structure at the front and a mechanically fastened subframe at the rear onto which the rear suspension is mounted. In the case of major frame damage the complete spaceframe can also be replaced. The choice of alloys for extrusions and sheet is influenced by ease of recycling.

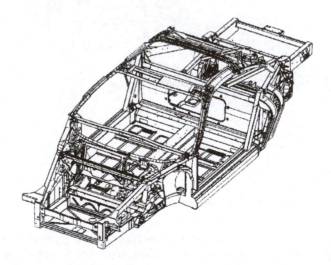

Fig. 2.26 Ferrari 360 Modena design

The Ferrari 360 Modena extends the aluminium spaceframe concept even further and comprises cast, extrusions and sheet.[12] A co-operative venture with Alcoa, the extruded and die cast components are made in Soest, Germany, a Ferrari supplier fabricates the sheet components and Ferrari supplies the sand castings including the integral parts of the spaceframe such as the front and rear shock towers. The spaceframe structure increases overall body stiffness (42 per cent in bending, 44 per cent in torsion) and safety while lowering the weight by 28 per cent and part count by 35 per cent compared to the steel predecessor. The F360 is claimed to be competitive in cost with a comparable steel body. This model is 10 per cent larger than the one it replaced. Materials used are summarized in Table 2.2.

Table 2.2 Ferrari 360 Modena materials

Alloy temper	Sand castings		Extrusions		Sheet components	
	B356-T6	CZ29-T6	6260-T6	6063-T6	6022-T4	6022-T6
0.2% proof stress (MPa)	170	125	200	160	130	275
UTS (MPa)	240	185	225	205	235	310
Elongation %	7	11	10	8	23	10

The spaceframe comprises 42 per cent extruded components and 33 per cent cast components, the remaining 25 per cent being formed sheet parts and stampings. All critical loads are transferred to the spaceframe through six castings. Sand casting was selected on the basis of low part volume and minimum weight requirements and these parts also provide significant part consolidation.

Most joining operations were carried out by MIG welding and self-piercing rivets with special emphasis on the achievement of extremely accurate build tolerances. Consistent conditions are maintained by using a machining centre for the location of reference locators. The final spaceframe is shown in Fig. 2.27.

2.7.4 The ASF aluminium spaceframe utilizing castings and profiles

Audi A8 and A2

A significant evolutionary step in the application of aluminium in autobody construction is the Audi A2, the first volume production vehicle to have body structure manufactured completely from aluminium. While the earlier 1994 A8(DT) version was clearly a major step forward in aluminium application, the spaceframe technology employed

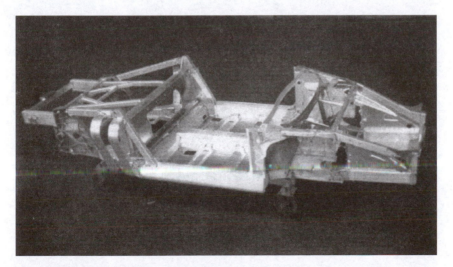

Fig. 2.27   Final 360 Modena spaceframe

was designed for medium volumes (A8 annual was typically 15 000 per annum). With the A2, aimed annual volumes are 60 to 70 000 units per annum, and the technology employed is more suitable for higher production rates. Other noticeable differences to the original Audi A8 is the increased use of complex castings and the widespread application of rolled aluminium profiles, Fig. 2.28. Interesting components included the 'B' pillar complex casting which replaces typically eight steel pressings in traditional vehicles and a world first application of laser welding aluminium (around 30 m in total).

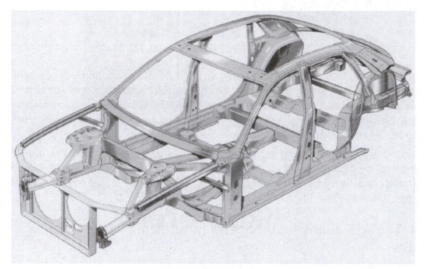

(a) 1994 (D2) version

(b) 2002 (D3) version

Fig. 2.28   Audi A8 body structure

Following a number of years of research, Audi unveiled the aluminium intensive vehicle concept Audi 100 in 100 per cent aluminium at the Hanover Fair in 1985. This development progressed to the ASF or Audi Space Frame design in 1987 and finally the Audi A8 production model. The frame structure was formed from straight and curved box extruded sections joined into complex die cast components at highly stressed cornered connection points. The load bearing parts are integrated as a structure mainly through the MIG welding process, with stressed skin panels attached mainly by the punch riveting process. This was one of the first applications of such a process in the automotive industry and one of the main reasons for this was the 30 per cent higher strength of joints made using punched rivets compared to spot welding. Resistance spot welding was used for joints, which were not accessible for punch riveting. The final assembly of the body structure illustrates the three major differences between the ASF concept and traditional steel monocoque construction:

- Fabricated spaceframe with extrusions and castings
- Manufacture of hang-on parts (closures) including extrusions for stiffeners
- Combining the separate front and rear body sections to form the final body shell

At the time of release by Audi, the ASF was claimed to exceed the rigidity and safety levels of modern steel bodies while achieving a weight reduction of the order of 40 per cent.

In many ways it is likely that in future years the Audi A8 will be regarded as one of the key technical developments in autobody materials technology. However, six years later in 2000, Audi unveiled the next stage of their aluminium body development, perhaps the more important A2, Fig. 2.29.

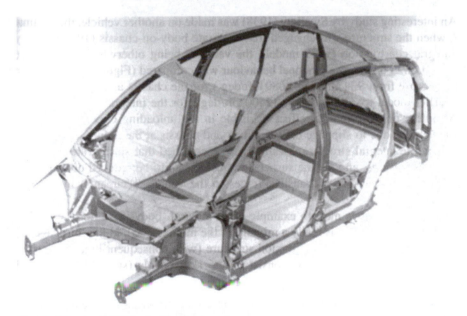

Fig. 2.29 Audi A2 body structure

While the original A8 was largely a hand-built car (a strategy that is acceptable for a production volume of 15 000 cars per year), the A2 was always intended to sell four times this number. This demanded a manufacturing concept integrating faster automated systems and techniques. The resulting A2 body structure is a highly innovative design taking elements of the A8's earlier concept but refining and adding technologies to them. In addition, the number of components has been reduced from 334 in the A8 to 225 in the A2. An excellent example of this part integration is the 'B' post component which in the A8 consisted of eight individual parts (extrusions, sheet, castings) integrated into one component, while the A2 'B' post consists of a single casting, Fig. 2.30.

Fig. 2.30   Audi A2 B post casting

The whole structure consists of 22 wt per cent aluminium cast elements, 18 wt per cent aluminium extrusions and 60 wt per cent aluminium sheet. The joining technologies used in the A8 have been refined for the A2. Spot welding and clinching were abandoned. The use of laser welding is of particular note, especially the floor pan laser welded to the spaceframe structure of extruded sections and pressure die casting. In total 30 m of laser welding is defined and the need for only one-sided access provides designers with extra styling freedom at the early concept stage of development. The most difficult aspect of laser welding is the tight tolerances for panel matching that are required (typically ± 0.2 mm). Compared with the A8, the self-pierce riveting process has been used increasingly to join sheet metal and extrusions.

In the latest version of the A8, announced in late 2002, much of the design and manufacturing technology has been carried over from the A2. However, it does represent a step change from the previous model. It is still essentially of ASF construction but the number of parts has fallen from 334 (including hang-on parts) to 267 through larger format pressings such as the side frame plus extruded sections such as the 3 metre long hydroformed roof frame, and multi-functional large castings used for the 'B' post (and radiator tank). The 'B' post previously comprised eight parts (4254 g) but is now a single component with the weight now reduced by 600 g. Compared with a conventional steel body weight has been reduced by 40 per cent. One hundred and fifty-six robots ensure an automation level of 80 per cent with a claimed 50 per cent saving in the production cycle and other manufacturing advances include a hybrid laser MIG welding process achieving synergistic effects by combining both joining processes. A hybrid welding seam length of 4.5 metres is achieved per body. As well as the hybrid welding seams there are also 2400 punch rivets, 64 metres of MIG welding seams and 20 metres of laser welding on each A8 body. Concerning the peripheral joint on bolt-on panels, rollers secured to a robot arm bend the outer panel over the inner and create a strong joint with the application of a hem-bonding adhesive. The doors, bonnet and tailgate are hemmed in this way and the wheel arch to side frame is similarly processed. Induction curing is used to prevent movement between inner and outer panels at the body-in-white stage.

The new A8 body is shown alongside the original version in Fig. 2.28 and a lot of development time and effort have obviously been expended in optimizing an extremely efficient production process. The emphasis on laser and MIG welding plus mechanical fastening appears to have advantages over the ASVT technology where adhesive application to structural joints must prove more of a challenge in terms of application and resulting consistency and durability.

2.7.5 Examples of hybrid material designs

2.7.5.1 Aston Martin Vanquish

The new Aston Martin Vanquish,[13] Fig. 2.31, contains a mixture of innovative materials technologies in a low volume model (350 units/year). A number of the skin panels are pressed or superplastically formed from aluminium sheet.

The body structure is mounted on an aluminium bonded and riveted lower structure (similar to the Lotus Elise), but incorporates a mixture of carbon fibre and aluminium extrusions in the floor/tunnel construction as shown in Fig. 2.32.

In addition to the tunnel the windscreen pillars are also carbon fibre bonded to the central structure to create a high strength safety cell. A steel, aluminium and carbon fibre subframe carries the engine, transmission and front suspension and is bolted directly to the front bulkhead. As can be seen above the doors are fabricated from aluminium incorporating extruded aluminium side impact beams.

2.8 Integration of materials into designs

2.8.1 General

The above appraisal of designs shows how the choice of material is paramount to achieving today's objectives and how this choice is widening. Having minimized the weight of a specific design and assuming the best materials have been specified, the

Fig. 2.31   Aston Martin Vanquish

Fig. 2.32   Internal structure of Vanquish showing carbon fibre composite and extrusion construction

next consideration is to optimize material with regard to each link in the chain of processing operations necessary to produce a functional part, and as will be described in Chapter 4 each of these elements can strongly influence the selection. For instance, complex parts require maximum formability which requires a compromise with strength, realistically placing a maximum at around 300 N/mm^2 proof stress, although for simple sections such as door reinforcement beams levels of 1200 N/mm^2 may be specified. The constraints imposed by local steelmakers may obviate certain grades where for instance a bake-hardening or isotropic steel is required, and a restricted choice of coating types may be available. However, despite these minor restrictions,

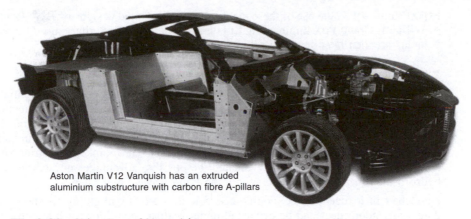

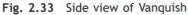
Aston Martin V12 Vanquish has an extruded
aluminium substructure with carbon fibre A-pillars

Fig. 2.33 Side view of Vanquish

apart from obvious exceptions most manufacturers are maintaining a conservative steel grade policy, requiring only minimal changes in processes, and as has been seen above the use of predominantly aluminium structures is only evident by one or two of the more adventurous companies who can absorb the extra supply and manufacturing costs. The majority would still prefer the more cautious approach employing the advantages of aluminium for closure or 'bolt-on' parts and using the accompanying weight savings to satisfy legislative weight-band requirements or added sports car performance. Many manufacturers are, however, gaining valuable manufacturing experience by building low volume sports models in aluminium, e.g. NSX or BMW Z8 or specific parts Peugeot 607. Once the different disciplines demanded by this less robust material are fully understood and a way is found of absorbing the extra cost it may then find a wider usage. Plastics as referred to later in this section require much development in an engineering context and only very expensive derivatives fulfil impact and other functional requirements. Until the market price falls then use will be limited to exterior cladding and trim items. Thus, for the main body structure the increasing use of high strength steel will continue to develop and the trend for a typically progressive European car manufacturer such as BMW is shown in Fig. 2.14 – a weight saving of 10–15 per cent being achieved for selective parts via thickness reduction.

Magnesium is now starting to find favour as the quest for lower density materials intensifies, but has been used occasionally in the past, e.g. Austin Maestro gear box covers. The latest interest is for vehicle cross-beams and similar aluminium products as described in the following sections.

2.8.2 Other materials used in body design

So far only primary materials have been considered but use now is being made of secondary forms of steel, aluminium and plastics. Examples are sandwich steel and similar aluminium products, described below, hydroformed steel and aluminium sections covered in Chapter 5, and tailor welded blanks (TWBs). High performance and competition cars are also making extensive use of honeycomb materials which when consolidated with composite skin layers provide ultralight high strength impact and structural sections. Because of their exceptional strength to weight ratio these may be the future first choice of body material for electric and alternatively fuelled cars. For an excellent introduction to Formula 1 body materials, which focuses on the use of carbon composites, the reader is directed to Chapter 4 which contains an expert summary on the use of these materials and associated designs by B. O'Rourke of Williams Grand Prix Engineering Ltd of Wantage, UK. Test criteria used for such vehicles are then later summarized in Chapter 8 by the same author.

2.8.2.1 Tube hydroforming

As evident in the ULSAB programme, hydroformed tube has significant potential in the consideration of parts consolidation especially for the more rugged applications such as the 4×4 sector which also allows a little more freedom of construction. The background and other weight-saving technologies demonstrated by the ULSAB initiative are gathered together in Chapter 4 but from a design viewpoint it is instructive to consider the recent study which evaluated the possible advantages in incorporating hydroformed structural elements within the Land Rover Freelander.

Described in a recent IBEC conference,[16] it is worth highlighting to show how potentially good ideas can be evaluated under realistic conditions, while defraying costs and resources of two major organizations. In this instance the design data for the recently developed Freelander was immediately to hand and could be relatively easily modified to allow an immediate comparison of new and conventional structures. The opportunity was also presented to allow full vehicle testing of a new concept rather than the body-only exercise with the ULSAB sedan, relying on FE modelling to predict performance.

The Land Rover Freelander was chosen as the focus for this programme principally due to the maturity of the development programme for the vehicle and the design package which allowed application to either smaller or larger products. Although a Land Rover (hitherto body-on-chassis design), the body is of monocoque or unitary construction, and the incorporation of a rigid sectional product seemed a natural choice for a rugged off-road performer.

The final configuration of hydroformed components incorporated in the design is shown in Fig. 2.34 and followed an extremely detailed study. It is worth mentioning that the normal procedure is to work to a controlled pre-development plan whereby the features of a new design are compared with the original, a cost-effective manufacturing route defined and rigorous testing of new components undertaken. The whole process is regulated with frequent timing reviews and concurrence obtained before proceeding through successive 'gateways' or decision points. These pre-concept stages constituting the 'creative' phase, gateways and process steps are illustrated in Fig. 2.35(a):

During this evaluation the advantages of the hydroformed sections will have been assessed, first, to confirm weight and space saving potential allowed by shape characteristics, Fig. 2.35(b) and second, as shown in Fig. 2.36 in comparison with other possible methods that could produce similar savings.

It was essential for comparisons involving joints and flange replacement that the welds were accurately modelled. For normal press steel box sections the assumption is made that flangeless sections are used and no allowance is made for distinguishing between alternative joining methods. However, it was critical for this new type of hydroformed joint that the joining method was represented more accurately and solid elements were used to represent adhesives and rigid bars positioned at the centre of flanges to simulate spot welds, Fig. 2.37.[10]

The various design iterations can then proceed to determine sections and joints which would probably benefit most from alternative hydroformed sections.

Comparison of hydroformed with conventional parts and the stages in the manufacture from tube are shown in Fig. 2.38.

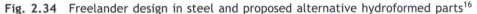

Fig. 2.34   Freelander design in steel and proposed alternative hydroformed parts[16]

The 'Application' phase comprised manufacture of the prototype parts illustrated in Fig. 2.38, using representative methods by a number of key tube hydroform suppliers, and finally a build and test programme to validate the advantages of the modified structure. The findings are summarized in Fig. 2.39.

The results shown in Fig. 2.39, and from crash and durability testing, demonstrated that the revised structure was equivalent in performance to the Freelander while torsional stiffness was markedly improved. However, starting a completely new model programme without the constraints of an existing body it is highly likely that far more significant weight savings and parts consolidation could have been achieved. Manufacturing feasibility was also demonstrated so opportunities can now be determined in the forward model programme.

2.8.2.2 Tailor welded blanks

The concept of producing composite blanks with tailored combinations of different thicknesses, strength grades and coated/uncoated steel provides the body engineer with the option of localized property variations wherever he wants them. Thus for longitudinal impact sections, controlled collapse may be induced as an alternative to 'bird beak' design, body side blanks may be blended to give formability in central areas and higher strength at pillar locations and door inner panels may be split to provide strengthening of the frontal area thereby dispensing with the need for reinforcement as illustrated in Fig. 2.40.

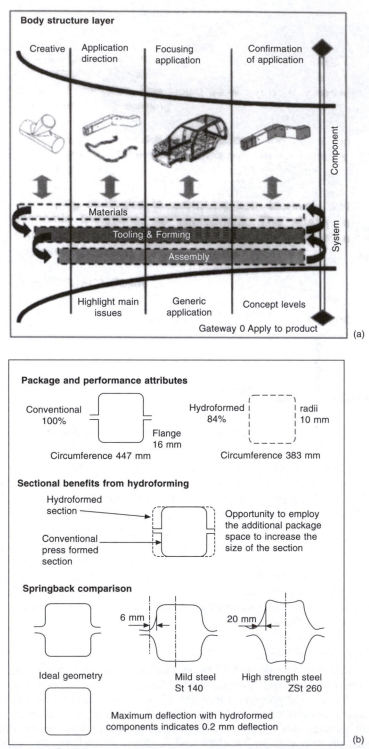

(a)

(b)

Fig. 2.35 Pre-development phases of ULSAB 40 and attributes of hydroformed sections.[16] (a) Pre-concept phases, (b) shape comparison with conventional sections. (Reprinted with permission from SAE paper 1999-01-3181 Copyright 1999 Society of Automotive Engineers Inc.)

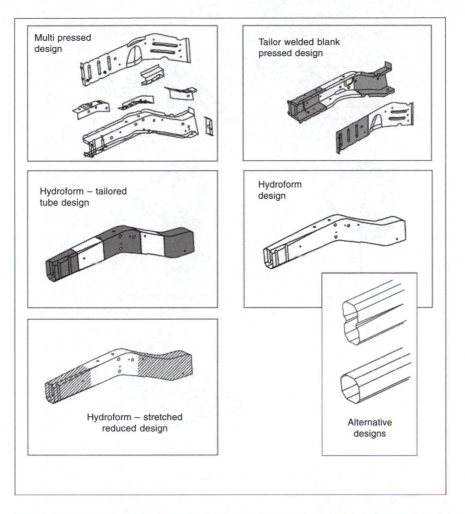

Fig. 2.36 Alternative forms of longitudinal section. (Reprinted with permission from SAE paper 1999-01-3181 Copyright 1999 Society of Automotive Engineers Inc.)

Thus increased scope exists for engineering solutions and parts consolidation which may offset the premium charged for the composite blank. This technology can be applied to steel or aluminium blanks. Composite steel blanks can be produced by mash or butt resistance welding but the finish containing a roughened fused weld zone is normally only suited to underbody parts. More often the blanks are now laser welded giving a narrow joint with a minimum of distortion and are widely used for most European models, typically for cross and longitudinal members, door inners and bodysides. Questions that must be addressed on order placement concern quality control procedures to ensure consistent weld quality, and liability in the case of failure of a structural part.

2.8.2.3 Sandwich materials

A material with extensive weight saving potential is sandwich steel. This consists of two thin sheet outers encapsulating a thicker polypropylene central layer. At present

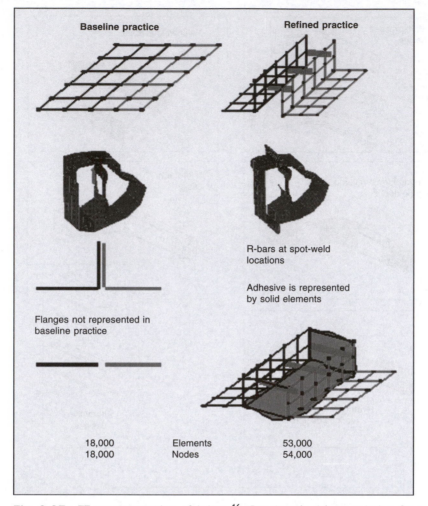

Fig. 2.37 FE representation of joints.[16] (Reprinted with permission from SAE paper 1999-01-3181 Copyright 1999 Society of Automotive Engineers Inc.)

there is not an extensive supplier base for these materials, since the commercial and engineering viability of the materials is not proven. Some of the versions that are on the market cannot resist the elevated temperatures during the body structure painting process. As a result, this material type is only viable for components that are assembled into the body after the painting process. In addition, this material is not weldable and must be assembled into the BIW by a cold joining process of either adhesive bonding or mechanical fastening. The ULSAB programme (Chapter 4) identified and subsequently defined two components in a sandwich steel material: a dash panel insert and spare wheel well. The steel skin used for the spare wheel well has a yield strength of 240 MPa and a thickness of only 0.14 mm. The core thickness was 0.65 mm, i.e. a total sheet thickness approaching 0.9 mm. The dash panel steel was a forming grade material (yield strength 140 MPa) with a thickness of 0.12 mm and a core of 0.65 mm.

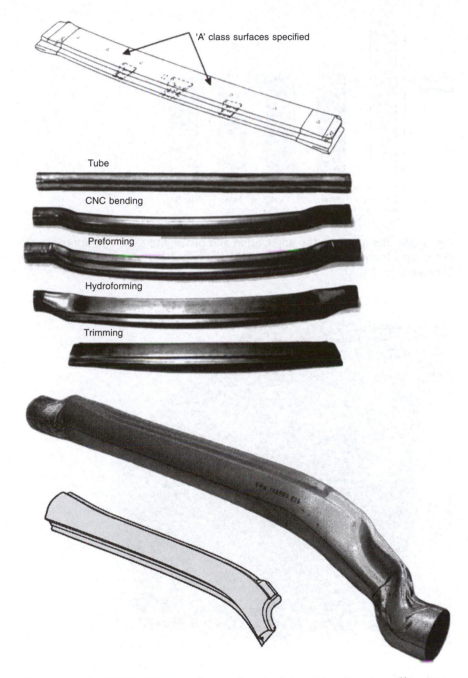

'A' class surfaces specified

Tube

CNC bending

Preforming

Hydroforming

Trimming

Fig. 2.38 Stages in component manufacture by tube hydroforming[16]

Even greater weight savings may be achievable through the use of an aluminium sheet version of the sandwich material. In this case typical thicknesses to achieve a similar level of bending stiffness to a steel panel would be 0.2 mm thick aluminium sheets surrounding a 0.8 mm thick thermoplastic core. Compared to steel this material offers weight saving opportunities even greater than aluminium, i.e. up to 68 per cent.

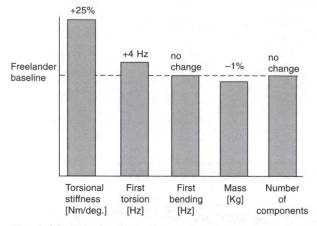

Fig. 2.39   Results from the application and proving phase.[16] (Reprinted with permission from SAE paper 1999-01-3181 Copyright 1999 Society of Automotive Engineers Inc.)

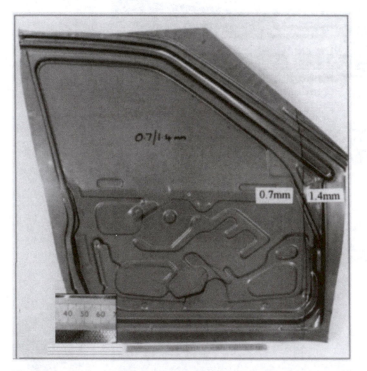

Fig 2.40   Tailor welded door inner blank showing thicker frontal area allowing deletion of separate reinforcement panel[17]

These sandwich materials could be argued to be good examples of the new type of hybrid materials technology that will be applied in the future, making use of the positive advantages of each material type, i.e. using the lightweight nature of the thermoplastic core and the stiffness, corrosion resistance and surface appearance of the metallic outer layers. However, application of this hybrid or composite material brings its own inherent difficulties with regard to recycling.

Customers now demand levels of in-car refinement that were unheard of a decade ago and one technique used within automotive design is to apply significant quantities of bitumen-based damping materials to critical regions of the body structure and closures. The main drawbacks associated with this approach are the additional mass and cost.

Laminated materials consist of two layers of conventional sheet material (usually steel) sandwiching a very thin layer of viscoelastic resin. The combination of these materials results in good sound damping performance (*x*). This material has been used previously in non-autobody applications, e.g. engine camshaft covers and oil sumps. Attention is now being focused on the application of these materials to panels such as the main floor and dash panels, which are typically covered in bitumen damping pads. Removal of these pads potentially offers weight and/or cost reduction opportunities along with NVH improvements.

Japanese motor manufacturers have pioneered the application of this material in body structures; examples of volume production use include the firewall panel on the Lexus LS400 and the Honda Legend. Clearly, there are many considerations about the use of this material through the process chain, including formability performance, welding and joining (see Chapter 6) and recyclability at the end of vehicle life.

2.9 Engineering requirements for plastic and composite components

The different types of plastic and respective manufacturing processes are referred to in Chapters 3 and 5 but reference should now be made to their engineering capabilities. Only very expensive derivatives such as carbon fibre composites fulfil impact and other essential structural needs. Until the market price falls the widespread use of polymers will be limited to exterior cladding and trim items.

Performance requirements for automotive body parts, and specifically plastics, are quite demanding. For example, vehicles must perform acceptably below −30°C, and under solar heating conditions, exterior components can reach temperatures in excess of 90°C. Panels must also be resistant to a wide range of chemicals and expected vehicle life can nowadays be in excess of 10 years or 100 000 miles during which material performance must be acceptable. These properties include:

- *Mechanical performance* – mechanical properties of relevance include tensile and shear strengths and modulus. Engineering thermoplastics typically have moduli of around 3 GPa and this relatively low value is related to the weak interchain bonds that hold the longer polymer chains together. In a thermoset where the chains are interlinked by strong chemical bonds, a higher modulus is exhibited (typically 4 to 5 GPa). Further increases can be achieved in composite materials through the addition of fibres, though the resultant modulus will still likely be less than that of metallic materials. As a result plastic and composite panels will usually be of greater thickness than metallic panels if a specific level of panel stiffness is required.

- *Impact* – impact performance (both low energy impacts or dent resistance and high energy in terms of crash performance) is a major consideration. To maintain polypropylene impact performance at low temperatures it is necessary to add additional components to the polymer blend to avoid brittle failure. For composites the fibre/matrix failure is the major energy absorption mechanism. In terms of dent resistance, polymeric panels deform in a different way to steel panels and many polymers can exhibit superior dent resistance performance by virtue of their low modulus. Material properties and their effect on dent resistance therefore become a prime consideration in panel design. The Land Rover Freelander 4 × 4 vehicle incorporates two new material applications for body panels, which as well as offering other benefits, provide improved dent resistance performance. On a vehicle designed for off-road use, the enhanced dent resistance of the plastic front fenders and zinc coated high strength steels should provide significant customer benefits.

- *Temperature performance* – high temperature performance in service is critical. Since polymers tend to have a greater rate of thermal expansion than steel, it is possible to have visual quality problems in terms of buckling, warping or uneven panel gaps. This expansion must be allowed for at the design stage – by appropriate design of the fixing method. Composites such as SMC have expansion rates more similar to steel and therefore this issue is of less concern.

- *Durability in-service* – both UV resistance and solvent resistance are key performance measures for exterior panels in particular. Unlike metallic panels, polymers can be susceptible to UV degradation and the addition of stabilizers is necessary to the base polymer. Solvent resistance is also critical, e.g. petrol, and again it is necessary to use a protective coating to ensure thermoplastic materials do not suffer a loss in strength or stiffness due to absorption of solvents.

2.10   Cost analysis

Many of the technologies described herein are aimed at achieving a reduction in component weight. Indeed, the selection of material type is based in part on careful consideration of the improved fuel economy derived from the use of lightweight materials versus the increased costs that are often incurred. This will be apparent from the table of selection criteria used by a prominent motor manufacturer in Chapter 3. Many different cost models can be applied to the evaluation of material types in these applications,[8] but general trends can be identified, Fig. 2.41.

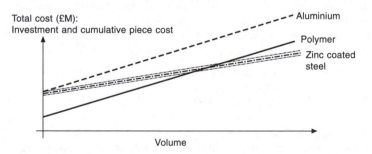

Fig. 2.41   General cost basis for automotive skin materials

The relative cost benefits/disadvantages of each material type are detailed for each stage of the process chain within each of the following chapters. However, since material selection is very often based on cost analysis at the design and engineering phase of the chain, an overall appreciation of the relative cost balance of the various materials is included here.

For a particular panel, this may result in an increased cost for plastic compared to zinc coated steel, when manufacturing in excess of a certain annual volume. This is because although tooling costs for plastics are lower than for zinc coated steel, raw material costs are higher. Thus, as total vehicle volume increases, the cost benefit derived from polymeric panels decreases until a certain break-even volume when steel becomes the most economical solution. This break-even volume is the subject of on-going debate, though is likely to be less than 200 000 cars. It is pertinent to note that most medium and high volume models involve the production of over 250 000 cars/year, explaining why use of plastics/aluminium has been mainly limited to low volume vehicles. Nonetheless, with improvements in technology, the cost advantage for polymeric panels may potentially shift to higher volumes, making the alternatives to zinc coated steel more attractive to the automotive industry.

The material costs must be considered only an approximate guide. Each material manufacturer will produce the common material grades at different cost levels depending on the exact specification of their production equipment. In addition geographical differences can exist, for example throughout the Eurozone (EZ) coatings have been considered to offer a cost advantage over galvanneal coatings in Germany while the reverse has traditionally been true in the UK. This may go some way to explain the pattern of material policy within European carmakers.

For a true comparison of the economics of body materials the input detail may also extend to include different design and manufacturing strategies. A more comprehensive cost analysis has been demonstrated by Dieffenbach.[15] First, he compares five different systems that could be employed to design and manufacture a mid-range sedan: steel and aluminium unibodies, steel and aluminium spaceframes, together with a composite structure, and a cost breakdown is shown in Table 2.3.

Again at low volumes costs reflect investment levels while at higher volumes material costs have a bigger influence. Thus the trends shown in Table 2.3 are mirrored in these studies with steel characterized by high investment cost, lower material and faster production rate. Conversely moulded plastic has a lower investment cost, higher material cost and slower production rate. The composite monocoque has the lowest cost up to about 30 000 vehicles per year; from 30 000 to 60 000 vehicles per year the steel spaceframe shows the lowest cost. For higher volumes the steel unibody shows the lowest cost. The aluminium spaceframes or unibody do not show a cost advantage although the aluminium spaceframe competes fairly well (a 15 per cent cost penalty) and compares with the steel unibody at high volumes. For outer panel assembly sets compression moulded SMC has the lowest cost for volumes up to about 100 000 sets per year above which steel has the lowest costs.

The challenges for the future for each category include lower tooling costs and scrap production (down to 25 per cent) for steel unibodies, lower raw material costs, e.g. by continuous casting, for aluminium unibodies, full exploitation of 40 per cent mass reduction potential apparently available from the steel spaceframe, while the aluminium spaceframe would benefit by the adoption of SMC (or similar) cladding (24 per cent cheaper than aluminium) which would make it cost competitive up to 80 000 vehicles per year. The composite monocoque is characterized by relatively expensive materials and clearly the challenge here is to reduce raw material costs, especially for carbon fibre composites.

A second approach proposed by Dieffenbach[15] is to use a stainless steel spaceframe clad with self-coloured composite panels where economies are proposed by deletion of various levels of the painting operation. This idea highlights another method of utilizing materials development to reduce costs, and the concept of pre-painted strip is introduced in Chapter 9, but it does pinpoint one target area which would produce massive savings and that is the paint shop. Costs presented for steel vs stainless steel are shown in Table 2.4.

Therefore comparing costs can be an extremely complex process requiring an intimate

Table 2.3 Body-in-white cost analysis presented by Dieffenbach[14]

Body-in-white cost analysis: key design inputs for selected case study alternatives

	Steel Unibody	Aluminium Unibody	Steel Spaceframe	Aluminium Spaceframe	Composite Monocoque
Geometry					
Overall vehicle mass (kg)	315	188	302	188	235
Mass as % of steel unibody	100%	60%	96%	60%	75%
Spot joints (#)	3250	3400		1000	n/a
Seam joints (cm)	n/a	n/a	4,000		6000
Piece count					
Total piece count (#)	204	224	137	137	41
Count as % of steel unibody	100%	110%	67%	67%	20%
Number of stampings	187	207	40	40	n/a
Number of castings	n/a	n/a	30	30	n/a
Number of roll/ hydroformings	n/a	n/a	50	n/a	n/a
Number of extrusions	n/a	n/a	n/a	50	n/a
Number of mouldings	n/a	n/a	n/a	n/a	7
Number of foam cores	n/a	n/a	n/a	n/a	34
Panels (inners/outers)	17	17	17	17	17
Materials					
Material prices ($/kg)	$0.77– $0.92	$3.00– $3.50	$0.77– $2.20	$2.00– $3.00	$3.13
Material density (g/cm³)	7.85	2.70	7.85	2.70	1.59

Body-in-white cost analysis: key fabrication input for selected case study alternatives

	Stamping	Casting	Hydro- forming	Extrusion	Moulding
Range of cycle times (s)	8-12	50-60	30-40	3-10	600-1,200
Range of labourers/fab'n line	4-6	2	2	2	2
Range of machine costs ($M)	$1.3-$7.5	$0.8-$1.5	$1.0-$2.0	$1.0-$2.0	$0.5-$1.0
Range of tool set costs ($M)	$0.2-$6.0	$0.1-$0.2	$0.1-$0.5	$3k-$7k	$0.1-$1.2

knowledge of the expected design and production scenario before accurate forecasts can even be attempted.

It is important to appreciate that the application of new material technologies as a means of vehicle weight reduction will usually often be decided by the vehicle programme development manager who may be willing to pay a cost penalty to reduce weight. This penalty may be influenced by the need for the vehicle to remain in a certain weight class or to move the vehicle into a lower weight class. For example, in the USA higher profit luxury vehicles have a negative rating on the company's CAFÉ rating. Production of a large number of heavy vehicles in this class may incur a cost penalty and the programme manager may decide that the cost penalty of introducing a new materials technology will be compensated for by the ultimate weight positioning of the final vehicle.

In conclusion the main evolutionary phases of the automotive body structure have

Table 2.4   Relative costs of steel unibody vs stainless
spaceframe[15]

The stainless steel spaceframe is found to have a cost
advantage of about $375 (23%) if paint is not included,
and $475 (30%) if paint is included

	Steel Unibody	Stainless Steel Spaceframe
Structure	$748	$522
Panels	$191	$191
Assembly	$261	$115
Paint	$415	$314
Total	$1,615	$1,142

been reviewed, and the role of materials introduced with respect to properties and
costs, and performance expected in service. We now move on to the production
processes for each of these likely materials to understand more fully the strengths
and weaknesses of each, enabling the exact specifications meeting design, process
chain, and environmental requirements to be met at minimum cost, both direct and
indirect.

2.11 Learning points from Chapter 2

1. Early chassis-based construction has now been replaced by body structures of
   unitary design. The spaceframe concept is increasingly popular allowing a mix of
   materials to be used, with ease of disassembly and repair.
2. Aluminium design using cast nodes, profiles and sheet has now been proven as a
   feasible design for volume production although material and vehicle insurance
   costs remain high.
3. FEM design techniques are now proving invaluable in reducing the timeframe of
   model development programmes. Parameters from a wide range of materials
   including high strength steels, aluminium and polymer variants can be used to
   help predict performance in dynamic situations, e.g. a crash. Lower strain rate
   programmes can help determine forming feasibility.
4. Contemporary design influences can introduce conflicting interests: ease of
   recyclability is not commensurate with increased use of plastics used to lighten
   body structure. There should be no threat to vehicle safety if larger, safer steel
   structures are gradually replaced by lighter alternative structures.
5. Specialized production techniques offering new forms of materials such as tailor
   welded blanks and hydroformed tube sections are allowing more freedom of
   design with opportunities for parts consolidation and weight reduction.
6. TWBs and use of lay-up techniques with composites such as carbon fibre now
   allow localized strengthening and stiffening of different body zones thereby shedding
   superfluous weight.
7. The combination of advanced composites and ultralightweight honeycomb structures
   could provide the basis for future alternatively fuelled vehicles as demonstrated
   by current high performance vehicles.
8. Polymers offer the designer undoubted advantages extending the range of body
   shapes and exhibiting good low speed impact, scuff and dent resistance. However,

the range of materials must be rationalized to allow simpler specification on drawings/electronic identification systems and ease the task of segregation for dismantlers.

References

1. BRITE EURAM II Low Weight Vehicle Project BE-5652 Contract No. BRE2-CT92-0264.
2. Brown, J.C., Robertson, A.J. and Serpento, S.T., *Motor Vehicle Structures*, Butterworth-Heinemann, 2002.
3. Davies, G.M., Walia, S. and Austin, M.D., 'The Application of Zinc Coated Steel in Future Automotive Body Structures', 5th International Conf. on Zinc Coated Steel Sheet, Birmingham, 1997.
4. Garrett, K., 'First Without Chassis?', *Car Design and Technology*, May 1992, pp. 56–61.
5. Hodkinson, R. and Fenton, J., *Lightweight Electric/Hybrid Vehicle Design*, SAE International, Butterworth-Heinemann, Oxford, 2001.
6. Ludke, B., 'Functional Design of a Lightweight Body-in-White. How to determine Body-in-White materials according to structural requirements'. VDI Berichte 1543 Symposium, 11 and 12 May 2000, Hamburg.
7. Ludke, B., 'Functional Design of a Lightweight Body-in-White for the new BMW generation', *Stahl und Eisen*, 119, 1999, No. 5, pp. 123–128.
8. Lewandowski, J., *Audi A8*, Delius Klasing Verlag, Bielefeld, 1995.
9. Muraoka, H., 'Development of an All-aluminium Body', *Journal of Materials Processing Technology*, 38, 1993, pp. 655–674.
10. Holt, D.J., 'Saturn: The Vehicle', *Automotive Engineering*, Nov. 1990, pp. 34–44.
11. Anon, 'Aluminium Spaceframe Makes the Lightest Lotus', *Materials World*, Dec. 1995, p. 584.
12. Novak, M. and Wenzel, H., 'Design Engineering and Production of the Alcoa Spaceframe for Ferrari's 360 Modena', SAE Paper 1999-01-3174.
13. Anon, 'The Aston Martin V12 Vanquish', *AutoTechnology*, Aug. 2001, Vol. No. 1, pp. 28–29.
14. Dieffenbach, J.R., 'Challenging Today's Stamped Steel Unibody: Assessing Prospects for Steel, Aluminium and Polymer Composites', IBEC '97 Proceedings, Stuttgart, Germany, 30 Sept – 2 Oct 1997, pp. 113–118.
15. Dieffenbach, J.R., 'Not the Delorean Revisited: An Assessment of the Stainless Steel Body-in-White', S.A.E. Paper 1999-01-3239.
16. Walia, S. *et al.*, 'The Engineering of a Body Structure with Hydroformed Components', IBEC Paper 1999-01-3181, 1999.
17. Davies, G.M. and Waddell, W., 'Laser Welding Allows Optimum Door Design', *Metal Bulletin Monthly*, July 1995, pp. 40–41.

3 Materials for consideration and use in automotive body structures

Objective: To review the choice of materials suitable for body manufacture and provide an understanding of salient manufacturing processes, product parameters and associated terminology. This is essential if the limitations of specific materials are to be appreciated and advantages/disadvantages of competing materials are to be compared prior to the design selection stage. A more detailed consideration of the significance of the properties required for the conversion of these base materials into components is given in Chapter 4.

Content: The range of materials that can be realistically considered for body structures is reviewed – the critical need for consistency of properties and effect on productivity is emphasized, together with the associated advantages of continuous production methods – key stages of both steel and aluminium manufacture are described – the significance of the final skin pass is explained and methods used to vary the texture of work rolls and strip surface are described – types and strengthening mechanisms of high strength steel (HSS) grades in yield stress range 180–1200 Mpa are graphically described and an introduction provided to the relevant polymer types and mode of manufacture.

3.1 Introduction

The main materials used in body construction are evident from the preceding chapter, and, as indicated, early choices were governed by the increasing needs of mass production and later, during post-war years, availability, as suppliers struggled to resume production. It will also be obvious that nowadays the choice has broadened considerably as materials technology has responded to the needs of the automotive engineer and that a far more enlightened understanding of materials parameters is required if this enhanced range of properties is to be exploited to the maximum advantage. The situation has advanced significantly from the days when 'mild steel sheet' was the universal answer to most body parts applications. As apparent from the 'Introduction' any distinction between grades was then generally made on the basis of formability and in the case of the few aluminium specifications, differentiation was by temper (O = annealed, H = hard, etc.).

The metallurgy of both steel and aluminium alloys has now advanced significantly offering a wide choice of mechanical and physical properties together with other attributes. The metallics choice can now be widened and it is also necessary to consider plastics, where 20 different types can be used within the motor vehicle. The traditional requirements of the body engineer have always been strength, in both static and dynamic terms, and elastic modulus, which governs stiffness/rigidity and imparts stability of shape. To these can now be added drawability and work-hardening parameters which are important with respect to forming and stretching respectively, the latter also having an important effect on energy absorption and impact resistance.

A number of surface parameters are now held to have considerable tribological (frictional) as well as cosmetic significance. Whereas the main requirement could be specified in terms of surface roughness, R_a, a range of 'deterministic' (pre-etched) rather than 'stochastic' (random, shot-blasted) finishes are now options. With the advent of CAD systems the design engineer can no longer rely on experience from previous models, or on a dedicated materials engineer for his materials choice. Instead he is often confronted with a series of predetermined choices, from which he must make his selection. For the correct choice at the engineering stage therefore it is important that the automotive engineer fully understands the parameters presented to him and their relevance in terms of production and application, and controlling metallurgical characteristics. A summary of key parameters is presented in Table 3.1 showing some basic properties and relevance to the automotive engineer. The stage of the strip production process critical to the development of these parameters is also identified together with other influencing factors.

Table 3.1 Key design parameters and relevant processing details

Parameter	Relevance	Influencing factors and *key processing stages*
Strength	Design	Imparted by composition, deformation and grain size; *alloying during smelting and mechanical and thermal treatment.*
Ductility	Forming, collapse characteristics	Lean composition and optimum heat treatment; *careful analysis and extended annealing cycle.*
Drawability index 'r' ('resistance to thinning')*	Press forming	Crystallographic texture requiring *optimum rolling and annealing schedules.*
Work hardenability 'n'*	Stretch forming Energy absorption	Composition and grain size dependent; *casting and rolling.*
Surface finish	Lubricity during forming Painted appearance	Imparted by roll finish *at temper rolling stage*

*Fully defined in Chapter 5.

The importance of strength, ductility and surface finish will be self-evident from the preceding text but the 'r' value provides a measure of the resistance of the material to thinning in the thickness plane during drawing via a favourable crystallographic texture. Likewise during stretching a high work hardening 'n' value spreads the strain over a more diffuse area thereby offsetting the tendency to form a local neck. A further essential consideration is cost and this is evident in Table 3.2 illustrating the type of data used by the design house of a key European vehicle manufacturer in its material selection process. The extended range of materials and interaction with engineering design parameters illustrates the wider vision of the contemporary designer together with his awareness of costs!

The key parameters and criteria applicable to the main structure and panels were defined by Ludke (ref. 7, Chapter 2) and were presented in Figs 2.17 and 2.18. In addition to the parameters of engineering significance it is now essential to consider ease of manufacturing in a much more detailed way as the stages representing the 'process chain' are each increasingly complex and require forethought as to the implications of the introduction of any new facet of materials change on productivity. Any responsible company now adopts a 'life cycle' approach to consideration of new materials and so to complete the selection process prior to the approval of a new material further criteria must be applied which rate acceptability according to emissions friendliness, ease of disposal and recyclability.

Table 3.2 Extended choice of materials and parameters used by a key car manufacturer. (Courtesy of B. Ludke, BMW Group)

Material	UK equivalent	1 E-Modul	2 Dichte	3 $\frac{E}{\rho}$	4 $\frac{\left(\frac{E}{\rho}\right)}{Preis}$	5 $\sqrt{E}$	6 $\frac{\sqrt{E}}{\rho}$
FeP04 St 14	DCO 4 Forming grade	210000	7,85	26752	22293	458,3	58,4
ZstE 300 P BH	HSS Bake hardening 300 Mpa YS grade	210000	7,85	26752	20578	458,3	58,4
S 420 MC ZstE 420 NbTi	HSS HSLA Grade 420 Mpa YS Grade	210000	7,85	26752	19816	458,3	58,4
BTR 165 VHF – Stahl	Ultra high strength steel 1100 MPa YS	210000	7,85	26752	19108	458,3	58,4
AlMg5Mn 10%kv	Aluminium–magnesium wrought sheet for internal parts	70000	2,70	25926	4321	264,6	98,0
AlSi1.2Mg0.4 10%kv, 190° c, 0.5hr	Aluminium–silicon skin panel material paint bake hardened	70000	2,70	25926	3704	264,6	98,0
AZ 91T6 Magnesium alloy	Heat treated magnesium alloy	45000	1,75	25714	4675	212,1	121,2
TiAl6V4 F89 Titanium alloy	Titanium alloy for automotive consideration	110000	4,50	24444	349	331,7	73,7
Kiefer – longitudinal	Pinewood grain longitudinal	12000	0,50	24000	6000	109,5	219,1
,, transverse	grain transverse	12000	0,50	24000	6000	109,5	219,1
Al$_2$O$_3$ (Keramic, massiv) 'spröde'	Fused alumina	370000	3,85	96104	481	608,3	158,0
GFK 55% force parallel to fibre	Glass reinforced plastic long. fibres	40000	1,95	20513	2051	200,0	102,6
,, ,, normal to fibre	transverse fibres	12000	1,95	6154	615	109,5	56,2
AFK 55%, TM – Typ parallel to fibres	Aramid fibre reinforced epoxy – long	70000	1,35	51852	519	264,6	196,0
,, ,, ,, – transverse	,, ,, – transverse	6000	1,35	4444	44	77,5	57,4
CFK 55% force parallel to fibre	Carbon fibre reinforced epoxy – long	110000	1,40	78571	1310	331,7	236,9
CFK 55% force normal to fibre	transverse fibres	8000	1,40	5714	95	89,4	63,9
GF-PA-12 (54%) parallel to fibre	Glass reinforced polyester 54% – long	35400	1,70	20824	1041	188,1	110,7
GF-PA-12 (54%) normal to fibre	54% – transverse	4400	1,70	2588	129	66,3	39,0
Glas (massiv) 'spröde'	Fused glass	70000	2,50	28000	18667	264,6	105,8

Hinweis: grau hinterlegte Materialien sind Basis für Anlage 2 bis 5

| | = parallel zur Faser

_|_ = quer zur Faser

Anlage 1: Materialeigenschaften

7	8	9	10	11	12	13	14	15	16	17	18	19
$\dfrac{\sqrt{E}}{\rho}$	$\sqrt[3]{E}$	$\dfrac{\sqrt[3]{E}}{\rho}$	$\left(\dfrac{\sqrt[3]{E}}{\rho}\right)$	$R_{p0,2}$	$\dfrac{R_{p0,2}}{\rho}$	$\left(\dfrac{R_{p0,2}}{\rho}\right)$	$\dfrac{\sqrt{R_{p0,2}}}{\rho}$	$\dfrac{\sqrt{R_{p0,2}}}{\rho}$	$\left(\dfrac{\sqrt{R_{p0,2}}}{\rho}\right)$	$10^{-6} \cdot K^{-1}$	A_5, A_{80}	Halbzeug Preis in
Preis			Preis			Preis			Preis			DM/kg (1996)
48,6	59,4	7,6	6,3	185,0	23,6	19,6	13,6	1,7	1,4	11,0	>40	1,20
44,9	59,4	7,6	5,8	340,0	43,3	33,3	18,4	2,3	1,8	11,0	>28	1,30
43,2	59,4	7,6	5,6	480,0	61,1	45,3	21,9	2,8	2,1	11,0	>20	1,35
41,7	59,4	7,6	5,4	1100,0	140,1	100,1	33,2	4,2	3,0	11,0	10(A5)	1,40
16,3	41,2	15,3	2,5	185,0	68,5	11,4	13,6	5,0	0,8	23,8	20(A5)	6,00
14,0	41,2	15,3	2,2	260,0	96,3	13,8	16,1	6,0	0,9	23,4	15(A5)	7,00
22,0	35,6	20,3	3,7	200,0	114,3	20,8	14,1	8,1	1,5	26,0	7(A5)	5,50
1,1	47,9	10,6	0,2	820,0	182,2	2,6	28,6	6,4	0,1	9,0	5(A5)	70,00
54,8	22,9	45,8	11,4	100,0	200,0	50,0	10,0	20,0	5,0	7,0		4,00
54,8	22,9	45,8	11,4	3,0	6,0	1,5	1,7	3,5	0,9	40,0		
0,8	71,8	18,6	0,1	500,0	129,9	0,6	22,4	5,8	0,0	8,6	0,0	200,00
10,3	34,2	17,5	1,8	950,0	487,2	48,7	30,8	15,8	1,6	6,0	2,0	10,00
5,6	22,9	11,7	1,2	475,0	243,6	24,4	21,8	11,2	1,1	6,0	2,0	10,00
2,0	41,2	30,5	0,3	1500,0	1111,1	11,1	38,7	28,7	0,3	-3,0	2,0	100,00
0,6	18,2	13,5	0,1	750,0	555,6	5,6	27,4	20,3	0,2	-3,0	2,0	100,00
3,9	47,9	34,2	0,6	1100,0	785,7	13,1	33,2	23,7	0,4	0,0	1,0	60,00
1,1	20,0	14,3	0,2	700,0	500,0	8,3	26,5	18,9	0,3	0,0	1,0	60,00
5,5	32,8	19,3	1,0	600,0	352,9	17,6	24,5	14,4	0,7	5,0	2,0	20,00
2,0	16,4	9,6	0,5	65,0	38,2	1,9	8,1	4,7	0,2	5,0	2,0	20,00
70,6	41,2	16,5	11,0	1000,0	400,0	266,7	31,6	12,6	8,4	5,0	3,3	1,5

To illustrate this process a listing of realistic main contenders is presented below together with an impression of relevant ratings. This is not meant to be a definitive presentation, and due to differences in local references some minor anomalies will be noted regarding values presented for material properties compared with Table 3.2, but it does offer a concise summary of the wide spectrum of factors governing the selection of materials today, possible current ratings and methodology adopted by the larger organization.

This text is not only intended for the design engineer but anyone involved in the process or supply chain whether in a technical or administrative capacity. It is in the author's experience that all those directly and indirectly involved with engineering and launch of automotive designs have benefited enormously by visiting suppliers and understanding the production route and terminology associated with the material they are using. The relevance of process features and tolerances applicable to certain aspects of the product can then be realistically appreciated. It is therefore considered that some background detail to relevant processes is useful at this stage and this is presented in the ensuing chapter following the list of main contenders and current ratings. These are compiled from a volume car perspective and it is possible that the niche and specialist car sectors may reflect slightly different ratings. The content of this chapter may be biased to the main contender, steel, but prominent coverage is also given to aluminium and plastics, with reference to other materials as appropriate.

3.2 Material candidates and selection criteria

The range of body materials that can be considered for volume car body construction is shown in Table 3.3, and it will be apparent that the criteria used by a major manufacturer when considering a new design extend beyond the range of physical and mechanical properties on which selection was once based. Not all factors are shown and it is easy to subdivide any of the columns shown. As will be illustrated later, the legislative requirements concerning, for instance, emissions and end-of-life (ELV) disposal are now influencing the initial choice of material, and increasingly the process chain or successive stages of manufacture must be considered to ensure that minimum disruption is incurred which may have consequences in productivity and quality. Any allowances for new materials will have been thoroughly proven at the pre-development stage of production.

Steel is still the predominant material used for manufacture[1,2] and the generally high ratings levels shown under the 'Ease of manufacturing' column reflect the provision already made by the industry for compatible facilities. The lower ratings evident for processing of aluminium, for instance, do not indicate that newer materials are an inferior choice but are more indicative of the size of change and introduction of new practices necessary to accommodate them. However, changes are inevitable to ensure the different legislative requirements are met and Table 3.3 perhaps indicates the 'pain' necessary to implement these lighter, but sometimes problematic, alternative materials. Indicative figures are also included for polymeric materials frequently used in specialist car manufacture and carbon fibre composites used in competition vehicles.

3.2.1 Consistency: a prime requirement

Whatever the material with regard to the physical and mechanical properties, the one key requirement that is essential to maintain manufacturing productivity is consistency.

Table 3.3 Main criteria and ratings for realistic selection of automotive body materials

Material	Design parameters					Ease of manufacturing* ('process chain')			Environmental** 'friendliness'		Cost
Criteria	YS MPa	UTS MPa	A80 min%	E.Mod GPa	D g/cc	Forming	Joining	Paint	CO_2 + emissions	Disposal (ELV)	Forming steel = 1
1. Forming grade steel EN 10130 DCO4 + Z	140 min	270 min	40	210	7.87	8	9	9	7	9	1.0
2. HSS EN 10292 H300YD + Z	300 min	400 min	26	210	7.87	6	8	9	8	8.5	1.1
3. UHSS – martensitic	1050– 1250	1350– 1550 min	5	210	7.87	4	7	9	8	8.5	1.5
4. Aluminium 5xxx	110 min	240 min	23	69	2.69	6	5	8	9	9	4.0
5. Aluminium 6xxx	120 min	250 min	24	69	2.69	6	5	8	9	9	5.0
6. Magnesium sheet	160 min	240 min	7	45	1.75	4	4	7	9.5	6	4.0
7. Titanium sheet	880 min	924 min	5	110	4.50	6	5	7	9	6	60.0
8. GRP	950	400– 1800	<2.0	40	1.95	8	7	8	8	5	8.0
9. Carbon fibre composite	1100	1200– 2250	<2.0	120– 250	1.60– 1.90	8	7	8	9	5	50.0+

*Based on range 1 = difficult to process, 10 = few production problems
**Ease with which prevailing legislation can be met: 10 = without difficulty, 1 = extensive development required

Once a piece of equipment has been set to operate within a given range of compositional, mechanical and dimensional properties, to ensure maximum output it must run continuously without disruption and the need for persistently reset process variables such as press and welding settings. Of course the tightest tolerances must be held with regard to tooling and machine efficiency but uniformity of material characteristics is essential if the full benefits of these facilities are to be realized. This is particularly true of state-of-the-art automated tri-axis progression presses and multistation robotic welding equipment where due to the momentum of the system and parts in process any delay due to defective processing can quickly lead to large quantities of scrap (or parts needing rework) being generated before the fault is detected. Rigid monitoring of the feedstock production process is therefore required with regard to both constituents and process, whereas for automotive assembly manufacture, statistical process control should be rigorously implemented, with all contributors to the supply chain demonstrating compliance with the requirements of BSEN ISO 9002, QS9000 and individual company quality approval procedures. Uniformity and the highest possible quality are the twin aims and continuous processing must be encouraged as batch manufacture of primary material, even with the best available controls, is by definition liable to local inhomogeneities in composition, temperature and other influences. Continuous processing, due to the scale of operation and improved operating efficiencies, must therefore be beneficial and although material might deviate slightly from the original specification, it can be utilized efficiently. With regard to steel manufacture, continuous casting and annealing processes have already been widely introduced (the advantages of which to the automotive industry are described later in this chapter), and casting to thin section dimensions is well on the way to reality. It is now interesting to note the coupling together of increasingly complex major operations as continuous processes.

Other more ambitious attempts have been made at amalgamation of pickling, rolling and annealing processes but are hampered by capacity imbalances. More widely, similar processing is being introduced in the aluminium industry where continuous annealing using rapid induction heating has been developed and continuous casting has been accepted practice for many years.

Uniformity of processing and product is therefore as significant as property levels and this will be the underlying theme throughout the following text. Before considering effect of materials in the context of design, manufacturing and service, however, it is essential that the manufacturing processes associated with each of the main materials is understood so that the implications of various grades, treatments and finishes can be appreciated in later sections.

As stated the individual materials are presented in order of the extent of utilization within the industry, commencing with steel.

3.2.2 Steel

Despite the quest for alternative materials, and the considerable amount of research that has been carried out to develop lighterweight materials in the last 30 years, most owners of the current generation of cars are driving an essentially steel structure which required approximately half a tonne of flat rolled steel product to manufacture. As described almost 15 years ago,[2] this material has exceptional versatility in terms of formability, strength and cost, and the industry has responded quickly to recognize the changing engineering needs arising from legislative and environmental requirements. Put simply, advantages of steel as an autobody material include:

• Low cost

- Ease of forming
- Consistency of supply
- Corrosion resistance with zinc coatings
- Ease of joining
- Recyclable
- Good crash energy absorption

The main disadvantages of steel in autobody applications are:

- Heavier than alternative materials
- Corrosion if uncoated

However, both these factors have been addressed over the last 20 years through the development of a much wider range of sheet and strip products. Higher strength steels with a wide range of yield strength values – now extending to 1200 Mpa – can now be supplied and, as will be seen later, designs can be suitably modified to either improve performance at existing thicknesses or downgauge with strength related parts. Although stiffness remains unaltered it is possible to offset decreased torsional rigidity, for example, by the application of structural adhesive in flange areas or elongated laser welded seams. The full range of steels used in automotive design, from the forming grades with a minimum yield of 140 N/mm² to ultra-high strength steels with values up to 1200 N/mm², is shown later in this chapter (Table 3.8) but an indication of types and properties is evident from Fig. 3.1.

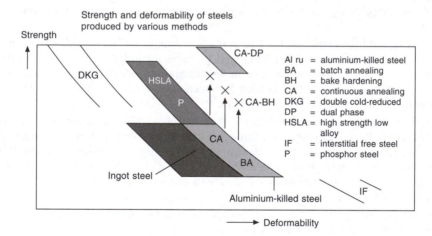

Fig. 3.1   Range of flat rolled steel products. (Courtesy of Corus)

At the lower end of the scale it is now possible to utilize very low carbon interstitial free 'IF' steels for some parts requiring exceptional deep drawing properties but it should be remembered that these are still very much the exception, and at yield/proof stress levels down to 120 N/mm² can depart from the accepted design minimum of 140 N/mm².

Likewise corrosion is much less of an issue than even 10 years ago. A range of zinc coated steels, namely, electrogalvanized, hot-dip, alloyed and duplex, is now available, the preference of individual automotive companies being dictated by cost, historical preference and manufacturing policy. The difference between the available forms, production processes associated with each of the coatings, and the mechanisms of corrosion and protection, are described fully in Chapter 7. Generally, the same range of steel properties is available with these coated products as for normal forming and high strength grades but sometimes a slight reduction in ductility is associated with hot-dip galvanized sheet due to the effect of the heat treatment cycle.

3.2.2.1 Steel production and finishing processes

Processing improvements which have enabled the increased range of properties previously highlighted are summarized in the following sections and many are highlighted in the flow chart shown in Fig. 3.2, the sequence existing at a major Corus installation, but typical of most plants worldwide manufacturing automotive strip.

Regarding steel production, it is probably sufficient to know at this stage that most steel used for autobody manufacture has been smelted from iron (produced in the blast furnace) and recycled scrap. Basic Oxygen Steelmaking (BOS) is the normal route whereby impurities are oxidized by the injection of oxygen through the bottom of the converter (included in Fig. 3.2), to produce refined material of composition typical of forming grades specified in Euronorm EN 10130. Normally these steels are aluminium killed (AK), an aluminium addition minimizing ageing effects by combining with nitrogen, and forming the characteristic 'pancake shaped', i.e. elongated, grains evident in the microstructure. Carbon levels are typically 0.03–0.05 per cent but for ultra-deep drawing coated or high strength grades where extra drawability is required, this is reduced to less than 0.0002 per cent. This is achieved by vacuum degassing of the molten steel prior to casting, resulting in the now well-known IF steels used for more complex shaped parts.

Vacuum degassing

This process involves the removal of gaseous and particulate inclusions, ensuring that very low levels of impurities are retained. Additions of titanium or niobium ensure interstitial elements such as carbon and nitrogen are reduced to extremely low levels and other compositional changes effected to optimize texture development, thus resulting in high 'r' values – hence the term 'interstitial free' and the associated high level of formability accompanying IF variants. This is an important treatment for high strength steels such as grades H180–260YD included in EN 10292 which show higher 'r' and 'n' values than other grades classified with similar strength levels. Similarly IF substrates can boost the formability of hot-dip galvanized products where annealing cycles might not be fully optimized. A typical vacuum degassing rig is shown in Fig. 3.3.

Continuous casting

Following the steelmaking process the molten steel is cast into a shape suitable for re-rolling. The traditional ingot casting processing route which led to the differentiation between 'rimming' and 'killed' steels has now largely disappeared as the continuous casting of slab has been introduced. With regard to ingot manufacture better utilization was achieved with rimming grades, as gas evolution in the final stages of solidification offset the familiar 'V'-shaped pipe associated with the killed grades, which was later cropped off. On re-rolling the rimming steel was, however, susceptible to strain ageing in storage. As a consequence the yield strength increased and the surface was prone to the appearance of 'secondary stretcher-strain' markings which could show through the painted finish. Now continuous slab production using the type of rig shown in Fig. 3.4 ensures that the maximum yield is now obtained and at least the same quality of material can be produced, but with a much higher level of consistency regarding cleanliness and property variation.

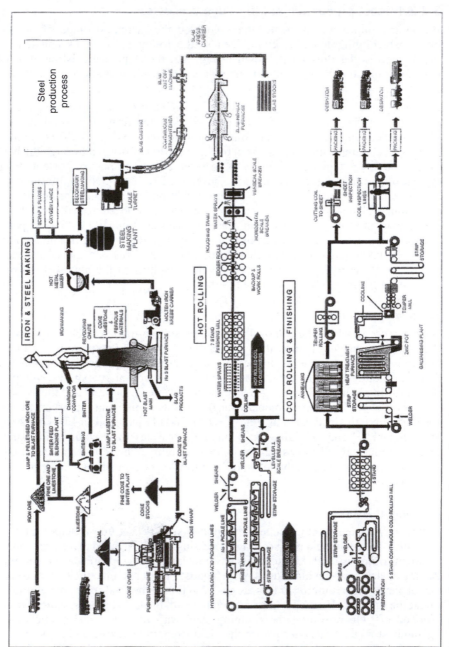

Fig. 3.2 (a) Typical sheet metal steel manufacturing process

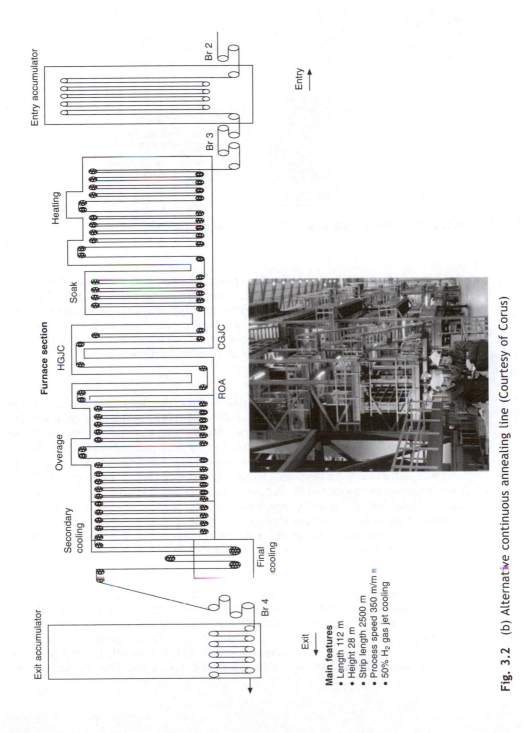

Fig. 3.2 (b) Alternative continuous annealing line (Courtesy of Corus)

Entry accumulator

Br 2

Entry

Br 3

Furnace section

Heating

Soak

HGJC

CGJC

ROA

Overage

Secondary cooling

Final cooling

Exit accumulator

Br 4

Exit

Main features
- Length 112 m
- Height 28 m
- Strip length 2500 m
- Process speed 350 m/m
- 50% H_2 gas jet cooling

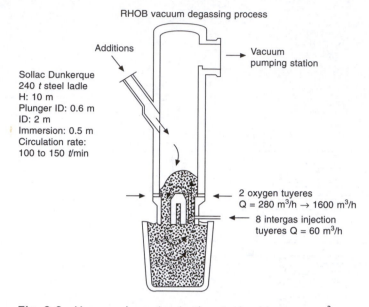

RHOB vacuum degassing process

Additions

Vacuum
pumping station

Sollac Dunkerque
240 *t* steel ladle
H: 10 m
Plunger ID: 0.6 m
ID: 2 m
Immersion: 0.5 m
Circulation rate:
100 to 150 *t*/min

2 oxygen tuyeres
Q = 280 m³/h → 1600 m³/h

8 intergas injection
tuyeres Q = 60 m³/h

Fig. 3.3 Vacuum degassing in the steelmaking process[3]

As shown in Fig. 3.4, the ladle of steel is poured directly into a water cooled copper mould, sticking being prevented by the reciprocating motion of the mould sides and a suitable dressing. The resulting slab thickness is typically around 250 mm but this has been reduced in some specialized 'mini mill' operations to 50 mm, thereby shortening the rolling process, and more recent advances aimed at direct strip manufacture make less than 2 mm material a realistic prospect.

Hot and cold rolling processes

After casting the slabs are progressive rolled down to the sheet thicknesses supplied in coil form to the automotive press shop. Following reheating the slabs are hot rolled to produce a 'hot band' at 900–1200°C, an intermediate form of now recrystallized material about 3 mm thick, but this has developed a thick layer of oxides or scale and although this is removed by 'pickling' in hydrochloric acid, the rough surface only renders it fit for selected underframe, chassis parts or bracketry. Process modifications have improved the quality of the hot rolled product and it is now possible to utilize this in thicknesses down to 1.6 mm, which gives a slight cost reduction to the part producer. Cold reduction is now essential to optimize dimensional accuracy, surface and properties, producing automotive material commonly 0.5–2.00 mm thick. This is normally carried out on a four- or five-stand sequence of four-high roll stations. The important influences on the final properties concerning formability are the grain size (controls 'n' value) and crystallographic texture which is developed on subsequent annealing – which in turn controls the 'r' value. Both parameters will be defined in detail in Chapter 4 and their importance in component manufacture explained. The hot rolling coiling temperature, percentage final reduction and annealing temperature/ rate critically affect these parameters, a high CR figure (of the order of 80 per cent) being commensurate with the optimum 'r' value. This is controlled by the power of the mill equipment, and the roll diameter/configuration and final thickness. The

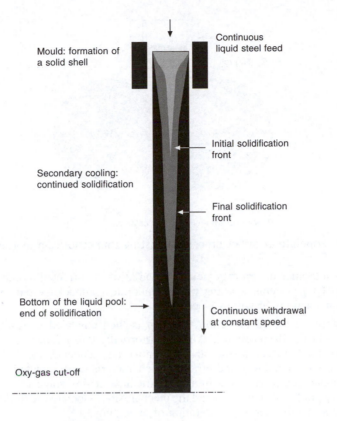

The principle of continuous casting

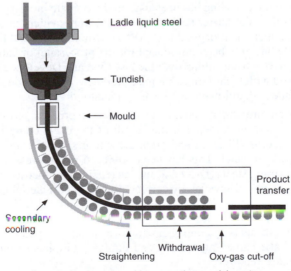

Layout of a curved mould type continuous slab caster.

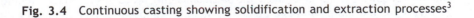

Fig. 3.4   Continuous casting showing solidification and extraction processes[3]

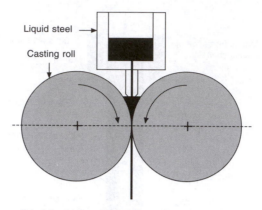

Liquid steel

Casting roll

Principle of twin-roll thin strip casting.

Fig. 3.5 Principle of direct process of casting thin strip from molten steel

thickness tolerance determines yield, and variation is corrected by automatic gauge control (AGC) systems working on an elaborate feedback system, which should achieve an accuracy of within 1 per cent.

Immediately after cold reduction the strip is then annealed to restore maximum ductility, and finally receives a skin pass (normally 0.8–1.2 per cent reduction) to impart the final surface texture and to remove any tendency for 'stretcher–strain' formation. The metallurgy and processing characteristics associated with this stage of steel manufacture are of extreme significance at this stage of manufacture of automotive grades of steel sheet and the mechanisms associated with skin passing are presented in detail after a consideration of annealing.

Continuous annealing

Although strand annealing has been used in the tinplate industry for annealing of thin strip,[4] and other similar research programmes were evident in the UK in the 1960s, continuous annealing process line (CAPL) technology (typical line illustrated in Fig. 3.2(b) p. 71) has only been introduced for the processing of automotive sheet, as an alternative to batch annealing, over the last 15 years. Developed extensively in Japan it is now used widely in Europe but the differences of the product compared with that of batch annealed coil need to be fully understood.

As suggested previously, advantages should accrue in terms of consistency of composition and properties, but the length of the cycle being less than 10 minutes means that recrystallization and grain growth are not as complete as for the batch annealed (BA) product. This has been observed in the press shop where for lower grades of forming steels (DCO 2/3) the 'as received' properties have not matched the BA ductility levels. However, by correct allocation of material to the less demanding jobs and a reasonable degree of rework to tools, the CAPL products are now in regular use. The rapid annealing capability has also been used to tailor the properties of high strength steels by control of composition and temperature/time cycle with the result that bake-hardening options and duplex/multiphase structured steels can be produced more readily.

Details of the differences between CAPL and BA processing are shown in Table 3.4.

Although requiring lower investment, the batch annealing processes commonly require 2–3 days for the more formable grades of steel as opposed to 10 minutes for the CAPL treatment, but even with the hydrogen atmosphere utilized by the Ebner process,

Table 3.4 Comparison of batch and continuous annealing process characteristics[1]

Grade	Temp. (°C)	Batch annealing of coils	Continuous annealing
		Slow heating cycle ≈ 30 ± 10 h Annealing temp. 710°C	Rapid heating cycle ≈ 90 ± 30 s 30 to 60 s hold Annealing temp. 850°C
	500–550	AlN precipitation Recrystallization	
	550–600	Grain growth	Start of primary
	600–650	Texture reinforcement	recrystallization
		Grain growth	Start of AlN precipitation
	650–700	Texture reinforcement	Grain growth impeded
Aluminium-killed extra mild steel with nitrogen in solution		Solutioning and partial spheroidizing of Fe_3C Renitriding in the case of annealing in HNX* (N tied up by excess Al)	End of primary recrystallization End of AlN precipitation,
	700–750	Start of secondary coarsening Coalescence of cementite Renitriding Loss of toughness (coarse grains)	grain growth impeded
	750–800		Grain growth impeded
	800–850		Very slow grain growth, start of spheroidizing of cementite
		25 ± 5 h	10 ± 5 mn
	700–600	Formation of Fe_3C nuclei	Formation of Fe_3C nuclei
	600–200	Complete precipitation of dissolved carbon	Partial precipitation of dissolved carbon Residual C 4 to 15 ppm depending on the overaging cycle

which provides a higher heat transfer efficiency, the uniformity of properties is still a problem.

The lower processing time, while providing a vastly increased throughput via CAPL (Corus furnace section shown in Fig. 3.2b) limits the development of favourable crystallographic textures and grain size, and as a consequence 'r' values are lower and drawability is impaired. Yield stress tends to be higher as it is related to grain size by the Petch equation[5] shown below,

$$\sigma_{LYS} = \sigma_i + k_y d^{-1/2}$$

where σ_i is the stress required to move free dislocations

2d is the grain diameter

k_y is the effect of dislocation locking by impurity atoms

Relative properties for each type are shown in Table 3.5.

However, as stated, through tool and press adjustment and possibly the use of an enhanced lubricity mill oil, similar performance can generally be achieved for the same part.

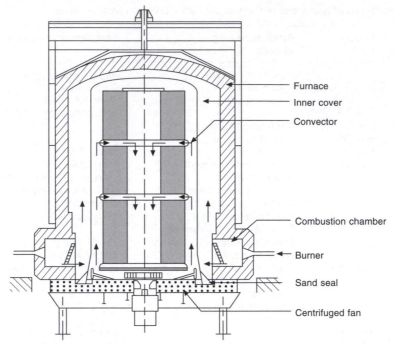

Schematic cross-section through a bell annealing furnace.

Fig. 3.6 Bell annealing furnace showing the vertical positioning of cold reduced coils

Table 3.5 Relative properties for batch/continuous annealed steel[3]

Batch annealing and cooling of coils	Continuous annealing and cooling
Final properties Grain size 7–9 ASTM YS: 160–190 MPa UTS: 290–315 MPa El.: 40–44% r: 1.6–2.1	Final properties Grain size 10–11 ASTM YS: 230–270 MPa UTS 330–400 MPa El.: 31–38% r: 1.0–1.4

Skin passing – effect on yield stress/yield point elongation

Unless IF technology is used during the steelmaking process, the presence of carbon and nitrogen interstitial atoms can lead to strain ageing on subsequent forming of the steel strip leading to the formation of stretcher-strain marks on the surface, accompanied by a well-defined yield point. This phenomenon is related to the pinning of residual dislocations by the interstitial atoms[6] and, simply stated, prior to skin passing the number of free dislocations is relatively few as they are locked on cooling from the annealing temperature. The 'locking atoms' are principally carbon, as any nitrogen has been combined with aluminium or an alternative addition, and these migrate to the interstices associated with the available dislocations. On straining a given amount, the stress required is that to move a few dislocations very quickly and involves a relatively high friction stress. Once these multiply on temper rolling the individual velocity and accompanying friction stress drops and the yield point falls to a lower

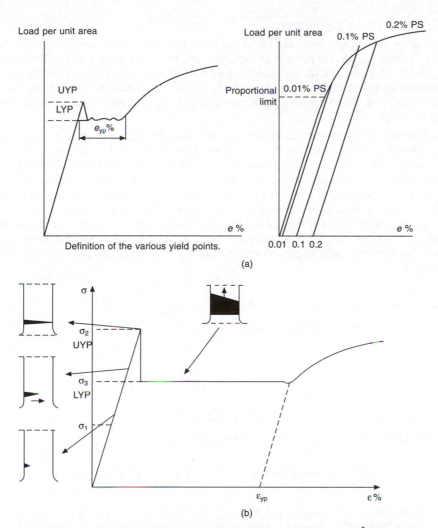

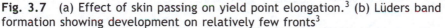

(b)

Fig. 3.7 (a) Effect of skin passing on yield point elongation.[3] (b) Lüders band formation showing development on relatively few fronts[3]

level. The yield drop (discontinuous yield point) can therefore be explained in terms of differential strain rate effects.[6,7] On a macro scale the initial yielding of a steel in the annealed condition takes place on only a few fronts and results in the formation of coarse Lüders bands as shown in Fig. 3.7(b).

This effect leads to a 'flamboyant' surface marking on press formed panels and for this reason the sheet is subjected to 'skin passing' after annealing. As described by Butler[7] this produces numerous blocks of alternatively deformed and undeformed material which result in a slight roughening of the surface, but acceptable for painting. When straining is resumed the velocity of each front and individual dislocation velocity is reduced, with an accompanying drop in friction stress. Therefore the process, also known as temper rolling, results in an overall reduction in yield stress with the coarse markings now subdivided on a much finer scale virtually invisible to the naked eye. Yielding starts at a much lower stress and the accompanying stress/ strain curve takes on a rounded smooth contour at lower strain levels, replacing discontinuous yielding as in Fig. 3.7(a).

3.2.2.2 Surface topography

As well as imparting a deformation of the order of 1 per cent to the strip to counter strain ageing effects, the process also dictates the final topography of the sheet. The type of finish embossed on the sheet surface by the work rolls that contact the sheet is becoming increasingly important as the lubrication characteristics during pressing and the finish developed during painting can both be optimized according to the final surface shape.

Traditionally the work rolls in the temper mill have been shot blasted and until the mid-1980s the optimum finish was defined in terms of R_a and peak density predominantly applied to a shot blasted texture. Other parameters were used in more detailed studies[8] such as Abbot curves but were difficult to apply on a routine basis especially in the workplace. Other surfaces from worldwide studies on sheet surfaces[9] examined using more sophisticated 3-D stylus plots of contour together with related parameters (skew, kurtosis) showed that the actual shape of the contours were important, plateau shapes resisting the collapse more associated with peaks (which can result in excessive debris formation), providing that channels were maintained to retain pressing lubricant.

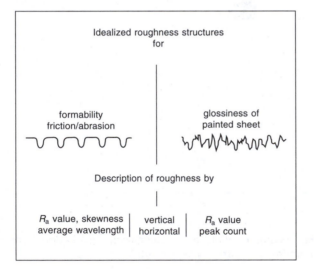

Fig. 3.8 Surface roughness profiles suited to forming and painting

For outer panels where maximum paint lustre was required a closer texture was needed in terms of peak spacing and the preferred finish as agreed by many automotive companies was as shown in Table 3.6.

Table 3.6 Typical surface texture finish for automotive panels

Application	Outer panels (µm)	Inner panels (µm)
Range 2.5 mm cut-off	R_a 1.0 to 1.7	R_a 1.0 to 1.8

In terms of consistency the shot blasted finish was not ideal as the actual surface contour varied with application mode and wear. Traditionally, a coarse texture from the tandem mill (main cold-reduction process) was required to avoid adjacent coil laps sticking on annealing and this was overlaid with the finer skin pass finish to provide the micro-texture producing the glossiness or lustre of the paint finish. The tandem mill texture was more open with peak spacings at 3–4 mm and relates to the coarse 'orange-peel' noted on many panel surfaces. With the advent of continuous annealing the tandem mill finish has less significance and the final finish is almost entirely due to the skin passing treatment. The control of texture using sand blasted temper rolls has always suffered due to lack of independent control over long and short wave contours and gradually development of alternative topographies has taken place over the last 30 years.

In 1971,[10] research had commenced on a more controlled system of surface preparation using electro-discharge texturing (EDT), the surface being eroded as in EDT machining by discharge of electrical energy through a di-electric, the work roll being one of the electrodes. As well as a more repeatable process the peak count can be double that achieved with equivalent shot blasted finishes.

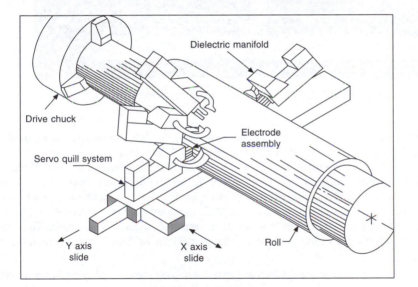

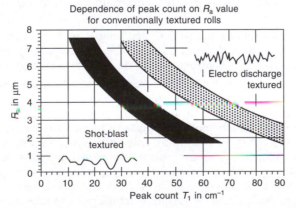

Fig. 3.9   Schematic diagram of an EDT system and comparison with shot blasting

In 1982, more deterministic or specific roll surface patterns were developed beginning with Lasertex which, as the name suggested, was produced by subjecting the work roll surface to intermittent exposure employing a mechanical chopper to cut the laser beam.

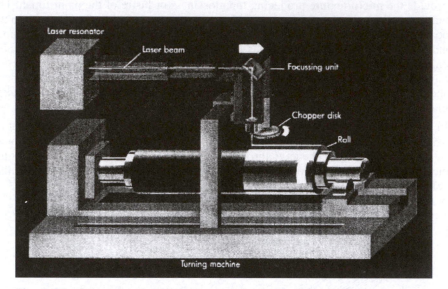

Fig. 3.10 Principle of the Lasertex method of roll texturing

This created a regular array of circular channels providing lubrication while maintaining a central core to resist local deformation/wear. While this was excellent for oil retention for deep drawn parts, the periodicity of this array tended to show through higher gloss paint finishes and fluctuations in application only imparted a semi-deterministic pattern. A 'mirror finish' version of this was produced on a much finer scale, which was claimed to enhance paint finishes by higher reflectivity from the flat fraction of the surface but has not been introduced on a wide scale. Thus the search continued for either a more regulated version or an alternative method of application.

The EBT finish obtained using electron beam technology and employing better beam/ workpiece synchronization results in a fully deterministic pattern as illustrated in Fig. 3.11.

The texture of the ERT rolls in the tandem and temper mill is generated by high energy and precisely positioned electron beam. The process takes place in a high vacuum chamber, a perfect atmosphere to avoid oxidation of the created crater surface.

The high performance electron beam is focused on the roll surface and melts the material. The locally created plasma blows the molten material aside, leaving behind a crater with a concentric rim (diameter from 50 to 250 µm, depth from 5 to 30 µm). The electron beam applies the homogeneous pattern to the total surface of the roll, which rotates at a constant speed (600 rpm). Due to the synchronization of electron beam and roll movement, the craters are generated on predefined positions on the roll surface. This technology leads to a complete and reproducible crater pattern. The total texturing process takes about 30 minutes. After the surface texturing process, the roll is returned to the mill and the texture is transferred onto the sheet surface at the final cold pass or temper reduction stage.

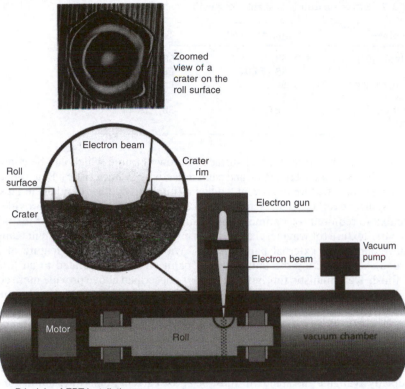

Zoomed view of a crater on the roll surface

Electron beam

Crater rim

Roll surface

Crater

Electron gun

Electron beam

Vacuum pump

Motor

Roll

vacuum chamber

Principle of EBT-installation

Fig. 3.11 Electron beam technology. (Courtesy Sidstahl)

The process can be applied to both the last roll of the tandem rolling operation and the temper mill work rolls. Some suppliers can now supply a range (e.g. the 'Sibertex' range) of finishes which are combinations of stochastic and deterministic topographies as shown schematically in Fig. 3.12, and a variety of finish combinations can be produced ranging from fully deterministic (regular) to stochastic (random) patterns as shown in Table 3.7.

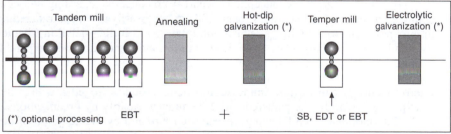

Tandem mill | Annealing | Hot-dip galvanization (*) | Temper mill | Electrolytic galvanization (*)

(*) optional processing   EBT   +   SB, EDT or EBT

Sibertex©-technology: working principle

Fig. 3.12 Tandem and temper mill treatments offered by the Sibertex process. (Courtesy of Sidstahl)

Table 3.7 Surface patterns available at specific manufacturing stages

Technology	Tandem roll	Temper roll
Shot blast (SB)	SB	SB
EDT	SB or EDT	EDT
Laser texturing	SB	Laser texturing
EBT-laser	SB	EBT
Simulation Sibetex©	EBT	SB, EDT or EBT

Thus, in summary, a full range of surfaces is now possible which can be controlled to give optimum lubricant retention and paint reflectivity characteristics. Batch annealing relied on a coarse texture to prevent 'stickers' (binding together of adjacent laps) and this was then overlaid with a finer texture on skin passing which provided the microtexture required for optimum paint finish. The interdependence of the tandem mill coarse texture (or waviness) and temper mill texture has to some extent hampered the development of external paint finishes. Waviness is a further feature of batch annealed sheet affecting the uniformity of paint and can be related to an 'orange-peel' finish. Deterministic finishes such as those described above provide more control over the final texture than random shot blasted treatment and for continuously annealed strip (no coarse texture for sticker separation necessary) should provide a uniformly fine and mirror gloss paint finish – if that is what is required. Many argue that the majority of customers might favour a 'chunkier' coarse texture indicative of a substantial paint presence. Either way methods of independently controlling the fine and coarse finishes now exist but for maximum benefit this steelmaking capability should be universally available and standards must be agreed at a national/international level to allow car manufacturers to specify the exact topography that best enhances their paint systems.

Effects in processing

As hinted at previously, the reason for modifying the surface is attributable to either benefits achieved in forming or influence over the final surface finish. BMW have carried out a very detailed investigation of surface on deep drawing and this is presented in Chapter 5, 'Component Manufacture'.

3.2.2.3 Higher strength steels

The application of higher stength steels in the automotive industry has been slow considering that basic grades have been available for at least 25 years, and this is due to a number of different preconceptions. Apart from concerns regarding the scope of application and the fact that high strength did not necessarily allow pro-rata reduction in thickness (due to the stiffness constraint, qualified in further detail in Chapter 2), formability was initially a key issue. Early grades, mainly rephosphorized and high strength low alloy (HSLA), were variable in properties and soon gained a reputation for tool wear and erratic performance, although as will be explained, process improvements have made these immeasurably better products. Together with exchange of experience worldwide, gathered and disseminated rapidly by organizations such as the International Deep Drawing Research Group, and by the adoption of design rules, harder wearing tool materials, surface treatments and compatible pressing facilities, these materials can be used with little extra effect. However, any attempt to run higher strength grades on existing or obsolete equipment will soon expose any weaknesses and result in high scrap rates/rework.

Table 3.8   High strength steel grades commonly available in Europe

Type	Range of yield stress MPa	Strengthening mechanism	Relevant standard
Low carbon mild steel sheet	140–180	Residual carbon, Mn, Si	EN 10130
Rephosphorized	180–300	Solid solution hardening	PrEN10xxxx EN 10292 (hot-dip zinc coated)
Isotropic	180–280	Si additions	PrEN10xxxx
Bake hardening	180–300	Strain age hardening	PrEN10xxxx EN 10292 (hot-dip zinc coated)
High strength low alloy	260–420	Grain refinement and precipitation hardening	PrEN10xxxx EN 10292 (hot-dip zinc coated)
Dual phase	450–600 (UTS)	Martensitic (hard) phase in ferritic ductile matrix	PrEn10xyz
TRIP steel	500–800	Transformation of retained austenite to martensite on deformation	PrEn10xyz
Complex and martensitic steels	800–1200	Bainitic/martensitic (hard) phases formed by controlled heat treatment	PrEn10xyz

The following pages provide a basic understanding of the different types of higher strength steels by considering the simple metallurgy of mild steel and then introducing different strengthening mechanisms responsible for other grades which can range up to yield strength values of 1200 Mpa. Most of these steels have related Euronorm or National Standards either published or in preparation.

The range of higher strength steels currently available is summarized in Table 3.8 which also highlights the strengthening mechanism and appropriate standard for that group of steels.

To further explain the strengthening modes associated with these steel grades a more graphical account is given in Table 3.9.

Thus it can be seen that a wide range of steels is currently available to the designer to help meet weight reduction targets and respond to challenges in safety engineering. The extent to which these special steels can be utilized and the changes necessary to accommodate them in manufacturing are considered in detail in Chapters 2, 4 and 5. It should be stated that the increased corrosion risks posed by downgauging these grades is more than offset by the use of zinc coated variants for most applications while also allowing vehicle durability targets to be maintained. Further detail on the types of coating employed and warranties in current use are also presented later in Chapter 7.

3.2.2.4   Stainless steel

As chromium is added to steels, the corrosion resistance increases due to the formation of a protective film of chromium oxide. The range and complexity of stainless steels is high and therefore detailed examination is outside the scope of this text. However,

Table 3.9 HSS strengthening mechanisms. (Illustrations courtesy of Thyssen Krupp Stahl)

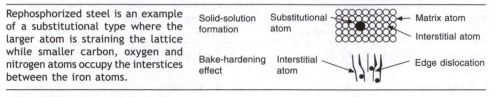

Rephosphorized steel is an example of a substitutional type where the larger atom is straining the lattice while smaller carbon, oxygen and nitrogen atoms occupy the interstices between the iron atoms.

Interstitial-free 'IF' steels are vacuum degassed to remove the carbon and oxygen atoms which impede the movement of dislocations and therefore increase the ease of deformation (positive effect on forming, negative effect on dent resistance). IF high strength steels therefore combine the increased ductility associated with the ferritic matrix but gain enhanced strength from substitutional phosphorus, silicon and manganese additions.

Bake-hardening steels derive their increase in strength from a strain ageing process that takes place on paint baking at circa 180°C. Sufficient carbon is retained in solution during either batch or continuous annealing to allow migration to dislocations following cold deformation. These are effectively locked, requiring a higher subsequent stress to recommence deformation, thereby increasing dent resistance.

The strength of both rephosphorized and IF high strength steels can be enhanced by this mechanism but different modes of carbon retention are required to prevent premature diffusion of carbon at either room temperature in storage, or during the application of zinc by hot dipping. In continuous processing this can be achieved by incorporating an overageing treatment and alloying whereby just enough carbon is retained in solution to allow the mechanism to occur at elevated temperatures.

The degree of cold deformation will reduce the ΔBH response correspondingly (see the diagram below).

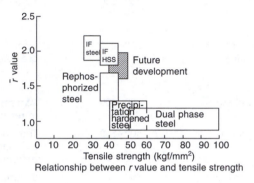

Relationship between r value and tensile strength

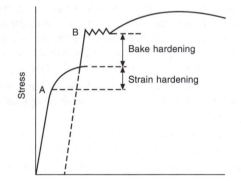

High strength low alloy steels gain their increased strength from the fine grain structure (smaller than ASTM No. 10) and fine dispersion of precipitates (e.g. niobium and titanium

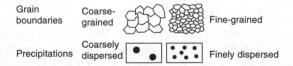

Table 3.9   *(Contd)*

carbo-nitrides) both of which impede dislocation movement thereby increasing the flow stress.

Multiphase steels derive their strength from thermo-mechanical processing, i.e. carefully balanced rolling, coiling and compositional control within the boundaries shown in the diagram opposite. Types of steel included in this category are dual phase, TRIP/TWIP, complex phase and martensitic phase as described below.

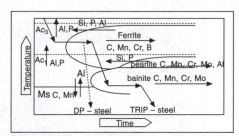

Influence of alloying elements on transformation behaviour

Dual-phase steels normally contain a matrix of ductile ferrite plus a proportion of the hard martensite phase induced by alloying and heat treatment. The characteristic high work-hardening rate results from the generation and piling up of dislocations around the martensite fraction on straining. The combination of the high strength developed, associated with relatively high elongation values, enlarge the area under the stress/strain curve resulting in improved energy absorption compared with other steels of similar strength. These steels also exhibit bake hardenability but unlike normal BH steels the ΔBH increase is not limited by the amount of cold work received (see below).

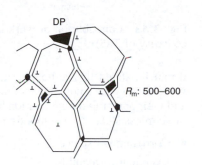

TRIP steels feature the transformation of metastable austenite to martensite during deformation thereby imparting a similar (but increased) strengthening compared to dual phase. The mechanism is similar to D-P with dislocation pile-ups at the martensite/ferrite phase boundaries.

(TWIP steels depend on the occurrence of mechanical twinning during deformation to achieve the necessary austenite phase change. These have a significantly different composition, e.g. 18% Mn, 3% Si and 3% Al and are under development for energy absorbing structural parts.)

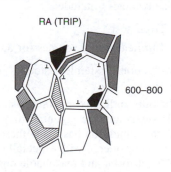

Complex steels are typically hot rolled, fine grain steels featuring ferrite, bainite and martensitic phases with a fine grained microstructure and uniformly dispersed superfine precipitates.

Martensitic steels are hot rolled with extremely high strength levels imparted by the predominantly martensitic phase.

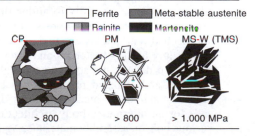

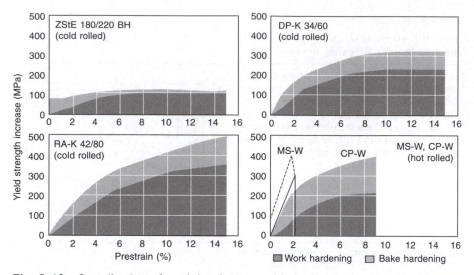

Fig. 3.13 Contribution of work hardening and bake hardening in HSS. (Diagram courtesy of TKS)

it can be stated that although stainless steels are not extensively used in current vehicles, they have nonetheless found applications in commercial vehicles, e.g. buses, and their potential for application has been advocated recently.[1] The main advantages of stainless steel as an autobody material are:

- Corrosion resistance
- Excellent formability
- Use of similar manufacturing infrastructure as mild steel

Disadvantages include:

- High cost
- Limited supply sources for automotive applications

Production of stainless steel is in many aspects similar to that described for mild steel.

The outstanding corrosion resistance of ferritic (13% Cr) and austenitic (18% Cr 8% Ni) steels and the potential they have for being used in the unpainted or partially painted condition has made them the subject of some very intense studies recently. When considered with especially attractive levels of forming parameters (Chapter 4) in sheet form, and reasonable on-costs, designers have been keen to re-examine the potential of this prestigious material which might give them a strong competitive advantage. It can be easily forgotten that these materials have high work-hardening rates and when considering press working with current facilities critical loads can be approached very quickly due to rapid work hardening. This behaviour can also demand heavy duty tooling materials and often interstage annealing to achieve deep drawn shapes, so despite the advantages of abbreviated paint processing overall costs for facilities and changed processes can be unfavourable. This assumes conventional pressed and hydroformed parts are used, but other recent studies[1] have made a strong case for an alternative body architecture utilizing a stainless steel spaceframe and various bolt-on assembly materials. Again reliance is placed on the need to paint only external surfaces.

3.3 Aluminium

In general terms, the attraction of aluminium is based on low density (2.69 g/cc) the relevance of which in automotive terms is discussed in Chapter 2. The historical rule-of-thumb when considering structures or subassemblies equivalent to steel is that the weight can be approximately halved but the cost is doubled. Although the density is one-third of steel, the full down weighting potential cannot be realized as the modulus (69 GPa) is considerably lower than that of steel (210 GPa), and as stiffness is a primary influence for the design of most body parts some compensation must be made and thickness increased. Any comment on cost must be qualified by the fact that this can fluctuate with the rise and fall of the commodity markets and for planning purposes some means of stabilizing future costs by buying ahead or alternative strategy must be considered. The doubling in net terms compared with the steel cost also includes a factor for increased manufacturing costs such as those incurred by modification of welding equipment, faster electrode tip wear, cold joining and the need for additional changes to the paint process. Total ownership costs must also be considered and one disadvantage of the more recent models featuring aluminium is the expense likely to be incurred for repair of specialist parts such as cast nodes which form part of a complex integrated substructure, and which could be reflected in high insurance cover – and the necessity to locate a specialist repair shop. To summarize the major advantages and disadvantages of aluminium as an autobody material:

Advantages:

- Low density
- Corrosion resistance
- Strong supply base
- Recyclability

Disadvantages:

- High and fluctuating cost
- Poorer formability than steel
- Less readily welded than steel

3.3.1 Production process

Aluminium is the most prolific metal comprising the earth's crust (8 per cent as opposed to 5 per cent for iron) but has only in the last 100 years been smelted industrially, the Hall–Heroult process being used to extract the metal from alumina dissolved in molten cryolite (a fluoride of sodium and aluminium) by electrolysis using carbon anodes. A flow chart showing the production process is shown in Fig. 3.14.

Following casting the slabs are milled to remove the tenacious oxide film and annealed for up to 8 hours at a rolling temperature of 440–550°C. They are then hot rolled to 10 mm then undergoing a continuous heat treatment before cold rolling finally on a four-high roll stand. The strip may then be straightened and cut to length.

3.3.2 Alloys for use in body structures

The common alloys used for the manufacture of body panels are shown in Table 3.10 and are designated according to the internationally recognized four-digit system. The

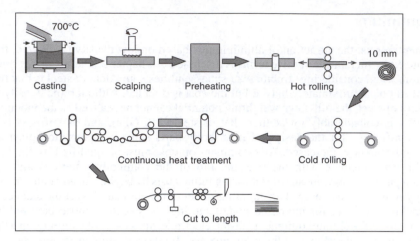

Fig. 3.14 Aluminium production process

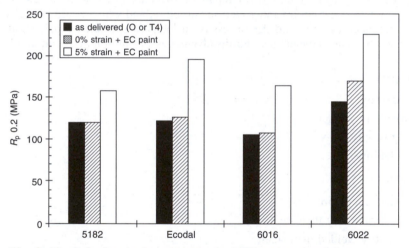

Fig. 3.15 Bake hardening response in 6000 series aluminium

5xxx series refers to aluminium–magnesium alloys while the 6xxx alloys refer to additions of magnesium plus silicon. The last two digits have no special significance beyond identifying different alloys; the second digit indicates alloy modifications and if zero indicates the original alloy.

The 5xxx 'wrought' series alloys have traditionally been used for panel production in the UK due to relatively low cost (3 × zinc coated steel, compared with 6xxx at 5 × zinc coated steel) and formability. The main concern has been that they are prone to stretcher strain markings or Lüders bands which can appear as flamboyant 'Type A' coarse markings on the sheet surface coincident with yielding, or as finer more regular 'Type B' markings which appear during the plastic stage of deformation. Despite claimed rolling and heat treatment solutions these tend to reappear on forming and can show through the paint finish unless reworked by abrasive discing.

The 6xxx series alloys are characterized by higher yield strength than Al–Mg alloys and are heat treatable imparting a significant degree of bake hardening at temperatures approaching 200°C.

Table 3.10 Automotive aluminium alloys in current use

Alloy AA / DIN	AA6016 AlMg0.4Si1.2	AA6111 AlMg0.7Si0.9Cu0.7	AA6009 AlMg0.5Si0.8CuMn	AA5251 AlMg2Mn0.3	AA5754 AlMg3	AA5182 AlMg5Mn
Temper	T4	T4	T4	H22 (Grade 3)	O / H111	O / H111
UTS (MPa)	210	290	250	190	215	270
0.2 proof stress (MPa)	105	160	130	120	110	140
Elongation A80 (%)	26	25	24	18	23	24
r (mean value)	0.61	0.55	0.64		0.70	0.80
n 5% (mean value)	0.30	0.28	0.29		0.35	0.33
Advantages	Formability, no stretcher-strain marks, balanced properties	No stretcher-strain marks, improved bake-hardening response	No stretcher-strain marks, mechanical strength	Corrosion resistance, cost	Good formability	Very good formability
Disadvantages	Limited bake-hardening response at Rover paint temperature	Corrosion concerns, limited formability	Limited hemming and forming properties		Possible stretcher-strain marks (Lüders lines) after deep drawing	
Alloy type		BAKE HARDENING			NON-BAKE HARDENING	
Typical use	SKIN PANELS			INNER PANELS		

Despite increased cost the 6xxx series alloys (6016 in particular) are proving most versatile and are in use by the majority of car producers using aluminium in Europe, providing a combination of good stretching and drawing characteristics, dent resistance and consistent surface. With regard to the latter, as well as being stretcher-strain free the use of a 1.0 micrometer EDT textured finish (see previous reference to surface finishes, p. 78) enables a similar quality of finish to be obtained as with outer panels in steel and this tends to be the universal specification even though advantages are being claimed for the EBT finish (see below). The other commonly used mill finish applied to internal panels and utility vehicles is less popular due to directionality effects on painting of vertical surfaces. For maximum economy, current designs often feature internal panels in 5xxx alloys with outers in 6xxx where critical quality is required.

Developments in alloys are summarized in Table 3.11 but include the emergence of an internal 6xxx quality, 6181A, and a 6022 alloy with a higher proof stress value than 6016 which may give further opportunities for downgauging providing forming and hemming performance can be sustained at realistic levels. It has been noted that the industry in the US has adopted the copper bearing 2036 alloy (not favoured in Europe for recycling reasons) for selected panels such as bonnets at gauges down to 0.8 mm compared with the more normal 1.2 mm in the UK and the potential for reducing thickness is now being explored with higher PS materials. Very high Al–Mg alloys (5.5% Mg content) are also being evaluated as elongation figures sometimes in excess of 30 per cent are achievable but high rolling load requirements make it very difficult to produce material with consistent properties.

As stated by Dieffenbach[1] the aluminium spaceframe represents the second leading body architecture and the history and current use of aluminium in design is presented in Chapter 2.

Table 3.11 Aluminium alloys under development

Alloy AA Rover/Alusuisse DIN	AA6022 (AlMg0.6Si1.3)	AA6181A EcodalR-608 (AlMg0.8Si0.9)	AA5022 (AlMg4.5Cu)	AA5023 (AlMg5.5Cu)	Pe-600
Temper	T4	T4	O/H111	O/H111	O/H111
UTS (MPa)	270	230	275	285	270
0.2 proof stress (MPa)	150	125	135	130	140
Elongation A80 (%)	26	24	28	29	29
r (mean value)	0.60	0.65	0.70	0.70	0.72
n 5% (mean value)	0.26	0.28	0.34	0.36	0.34
Advantages	Improved bake-hardening response	Improved bake-hardening response	Improved formability	Improved formability	Improved formability
Disadvantages	Directional hemming properties	Limited hemming properties	Corrosion, susceptible to stretcher-strain	Corrosion, susceptible to stretcher-strain	
Alloy type	BAKE HARDENING		SLIGHTLY BAKE HARDENING		NON-BAKE HARDENING
Typical use	SKIN PANELS		INNER PANELS		

3.4 Magnesium

Magnesium is the lightest of all the engineering metals, having a density of only 1.74 g/m^3. It is 35 per cent lighter than aluminium and over four times lighter than steel. It is produced through either the metallothermic reduction of magnesium oxide with silicon or the electrolysis of magnesium chloride melts from sea water. Each cubic metre of sea water contains approximately 1.3 kg of magnesium.

Common magnesium alloys are based on additions of magnesium, aluminium, manganese and zinc. Typical compositions and properties are shown in Table 3.12. The alloy designations are based on the following:

- The first two designatory letters indicate the principal alloying element (A for aluminium, E for rare earth element, H for thorium, K for zirconium, M for manganese, S for silicon, W for yttrium, Z for zinc).
- The two numbers indicate the percentages of these major alloying elements to the nearest percentage.
- A final letter indicates the number of the alloy with that particular principal alloying condition. Therefore, AZ91D is the fourth standardized 9% Al, 1% Zn alloy.

Table 3.12   Common automotive magnesium alloys

	AZ91	AM60	AM50
Composition			
% Al	9	6	5
% Zn	0.7		
% Mn	0.2	0.3	0.3
Typical RT properties			
UTS (MPa)	240	225	210
Yield strength (0.2% offset)	160	130	125
Fracture elongation	3	8	10

The higher elongation levels of the AM60 and AM50 alloys have meant that they may be preferred to AZ91.

High purity variants of these alloys with lower levels of heavy metal impurities (iron, copper and nickel) have vastly improved corrosion performance. The sand casting alloy AZ91C has now been largely replaced by its high purity variant AZ91E, which has a corrosion rate around 100 times better in salt-fog tests.

The major advantages of magnesium are:

- Very low density
- Ability to be thin cast
- Possible to integrate components in castings

Disadvantages include:

- Only viable as cast components (sheet and extruded magnesium not readily available)
- High cost at medium to high volumes

The main applications of magnesium alloys are discussed in Chapter 5.

3.5 Polymers and composites

3.5.1 Introduction

Polymers used for autobody applications may be split into thermoplastics and thermosets. Thermoplastics are high molecular weight materials that soften or melt on the application of heat. Thermoset processing requires the non-reversible conversion of a low molecular weight base resin to a polymerized structure. The resultant material cannot be remelted or reformed. Composites consist of two or more distinct materials that when combined together produce properties that are not achievable by the individual components of that composite. In autobody applications, reinforced plastics are the major composite material. For example, the term fibreglass consists of a plastic resin reinforced with a fibrous glass component. The resin acts to define the shape of the part, hold the fibres in place and protects them from the damage. The major basic advantages of composites are their relatively high strength and low weight, excellent corrosion resistance, thermal properties and dimensional stability. The strength of a polymer composite will increase with the percentage of fibrous material and is affected by fibre orientation. Tailoring the fibre orientation and concentration can therefore allow for strength increase in the particular region of a component.

3.5.2 Thermoplastics

Thermoplastics can be divided into amorphous and crystalline varieties. In amorphous forms the molecules are orientated randomly. Typical amorphous thermoplastics include polyphenylene oxide (PPO), polycarbonate (PC) and acrylonitrile butadiene styrene (ABS). Advantages of amorphous thermoplastics include:

• Relatively dimensionally stable
• Lower mould shrinkage than crystalline thermoplastics
• Potential for application for structural foams

Disadvantages include:

• Poor wear abrasion and repeated impact
• Poor fatigue resistance
• Increased process times compared to crystalline thermoplastics

In a crystalline variety there will be regions of regularly orientated molecules depending on factors such as the processing techniques used, cooling rate, etc. Examples include nylon (PA), polypropylene (PP) and polyethylene (PE). Advantages of crystalline thermoplastics include:

• Good solvent, fatigue and wear resistance
• Higher design strain than amorphous grades
• High temperature properties improved by fibre reinforcement

Disadvantages include:

• Potentially high and variable shrinkage
• Difficult to adhesive bond (see Chapter 6)
• Higher creep than amorphous thermoplastics

3.5.3 Thermosets

Thermosets are generally more brittle than thermoplastics so they are often used with fibre reinforcement of some type. Advantages of thermosets include:

- Lower sensitivity to temperature than thermoplastics
- Good dimensional stability
- Harder and more scratch resistant than thermoplastics

Disadvantages of thermosets include:

- Low toughness and strain at fracture
- Difficulties in recycling
- Difficult to obtain 'A' class finish

There is a wide range of different processing techniques that can be used to produce components from the above raw materials. The basic processes are described in Chapter 5, although a number of excellent texts exist to provide more detailed information.

As will become apparent in Chapter 8 the main problem with plastics concerns ELV (vehicle disposal). While the metallic content represents most of the 75 per cent recycled content, plastics are a main constituent of shredder fluff or ASR (auto shredder residue) which can only be disposed of by landfill, and until rationalization of the types of plastic used takes place, focused on materials which are easily recycled, non-preferred types will be filtered from the initial approval process at the life-cycle analysis stage.

3.5.4 Polymer and composite processing

There are a number of ways of processing thermoplastic materials for automotive applications including extrusion, blow moulding, compression moulding, vacuum forming and injection moulding. However, some of these processes are more directly applicable for the production of autobody structures and closure parts than interior and exterior trim parts.

3.5.4.1 Injection moulding

This is one of the commonest routes for producing thermoplastic components and has been used in a number of autobody applications including the plastic fenders on the Land Rover Freelander and Renault Clio and the vertical panels on the BMW Z1. The process involves feeding polymer granules into a heated extruder barrel which heats the compound, Fig. 3.16. The resultant melt is injected into a chilled mould and pressure is maintained during cooling. The part is finally ejected. Advantages of the

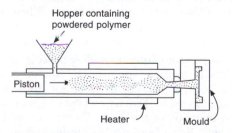

Fig. 3.16 Injection moulding process

process include the relatively short production times and the ability of produce complex, precision parts. However, the pressures required during injection are high and necessitate the use of a precision tool which leads to high tooling costs and lead times.

3.3.4.2   Glass mat thermoplastic (GMT) compression moulding

Glass mat thermoplastic (GMT) is produced in sheet form that is cut into blanks, rather like traditional sheet metal. These blanks are pre-heated prior to loading into cooled tools within vertical presses. The tool is closed under high pressure and the material flows into the tool cavities. The main problem with GMT is the inability to achieve a truly 'A' class finish meaning its application is limited to internal applications, the most common being the GMT front end on many European production vehicles. The advantages of the process include the faster cycle times than SMC moulding, consistent quality and the potential to use modified metal stamping infrastructure.

Thermoset processing includes sheet moulding compound (SMC), resin transfer moulding (RTM) and reaction injection moulding (RIM), which are illustrated in Chapter 5, and describe more specific aspects of component manufacture.

SMC compression moulding processes are similar to those described for GMT stamping. The SMC sheet is taken unheated and placed in a heated tool at around 160°C. This causes the resin to cross link and cure in the tool. The pressures are lower than those used for GMT and the resultant properties are higher with a modulus approaching twice that for GMT. Numerous SMC body panels have been used on production vehicles particularly in closure applications, including those from Ford, Lotus, Renault and Daimler Chrysler to reduce vehicle weight and investment cost. Bulk moulding compound is similar to SMC, but with bulk material replacing the sheet. The advantages of SMC include:

• Good surface finish possible
• Good accuracy of parts
• Viable for medium volumes

Limitations of the process include:

• Relatively high investment
• Not as significant weight saving as with thermoplastics
• Storage/shelf life of SMC

A more recent development in the field of SMC technology is the development of low density SMC. By replacing the calcium carbonate and other SMC fillers with hollow glass microspheres, the density of SMC can be reduced from the traditionals SMCs' gravity of 1.9 to as low as 1.3, with a small reduction in stiffness. These low density grades may be applicable to interior parts. In the longer term exterior grades may be possible but surface finish after repair is the major concern. When low density materials are sanded and repaired the hollow glass sphere can be opened up and development of an effective surface sealer is required before this can be resolved.

Resin transfer moulding (RTM) is a low pressure liquid moulding process. It has traditionally been used for parts with low to medium volume. The low pressures involved in the process allow the use of a low cost tool, one of the major advantages of the process. Fibre reinforcement is placed in the tool cavity and the tool is closed. Clamping pressure is applied before the injection of the resin. Cure times tend to range from a few minutes to many minutes. Because of cycle time limitations the process is only practically viable for volumes of up to approximately 40 000 units.

3.5.5 Advanced composites

The application of composite technology to Formula 1 racing cars will be described in Chapter 4 but additional detail on the technology is provided with reference to the F1 McLaren production car by Martin.[11] More than 95 per cent of the McLaren F1's body is constructed in high performance advanced carbon (graphite) epoxy composite material. The material starts life in the tacky pliable condition in which the fibres, in this case carbon, are embedded in a partially cured resin. The weight of the cloth and the resin held within it in the prepregnated condition are controlled to very tight limits. Currently material weights used vary from a 150 gsm (g/m^2) twill weave 1k high strength carbon fibre to 660 gsm 12k twill weave high strength carbon fibre. The 1 or 12k reference defines the number of carbon filaments that make up one strand or tow of the woven cloth, i.e. 12k means 12 thousand filaments. In this state it is workable and thickness, stiffness and strength of the final structure can be controlled to very fine limits. Laminating next takes place and after other specialized preparation processes, curing takes place in an autoclave programmed with two cure cycles. These are 125°C 2 bar (250°F at 30 psi) and 125°C 5 bar (250°F at 75 psi), and this phase takes 3 hours to complete. The advantage of this type of technology is that it allows the engineer to control the properties – including stiffness and strength in three dimensions – and develop these characteristics exactly where he wants them. Thus the maximum efficiency is obtained from each gram, i.e. the maximum structural and weight efficiency. The big advantage of this approach, apart from tailoring the properties in the required location, is that the strength to weight ratio is impressive and it is claimed that the same tensile strength as steel is obtained but at one-quarter of the weight. Unidirectional material in both the high strength and high modulus forms is used in specific areas where increased stiffness and reinforcement are required. This technique has allowed the impact resistance of carbon fibre composites evident in Formula 1 collisions to be transferred to production cars. As for the Formula 1 body described later, the lay-up technology can be used in conjunction with honeycombe panels to create a strong and stiff assembly.

SP Resin Infusion Technology (SPRINT)[12] provides an alternative approach which it is claimed is far less labour intensive than preimpregnated reinforcement, liquid resin infusion and resin film infusion techniques, while providing a higher integrity product with high quality external paint finish. The SPRINT material consists of two layers of dry fibre reinforcement either side of a precast, precatalysed resin film as shown in Fig. 3.17(a). The reinforcement can incorporate a wide variety of fibres such as glass, aramid and carbon, which can be in the form of random mats, woven fabrics or stitched fabrics. A number of resin systems such as epoxy, polyester and bismaelimide can also be used. Processing is straightforward with material laid up in a mould and vacuum bagged as for conventional prepreg. The nature of the product allows good consolidation and integrity, and flow of the material allows shape control, even into corners, without entrapment. Gel coats can be applied to give the desired finish. The comparative cost/weight effects are summarized in Fig. 3.17(b).

Specialist sports car manufacturers have already adopted this material and on the Ultima (shown in Fig. 3.18) – 45 kg has been saved compared to the GRP version. The SPRINT CBS material consists of a precombined series of materials with a surfacing ply on the outside, followed by a sandwich of carbon reinforcements either side of a thin syntactic core.

In conclusion, this appraisal of the process technology behind the manufacture of the main materials utilized in automotive body structures has allowed an understanding of the various aspects of process capability to be gained but also the development of

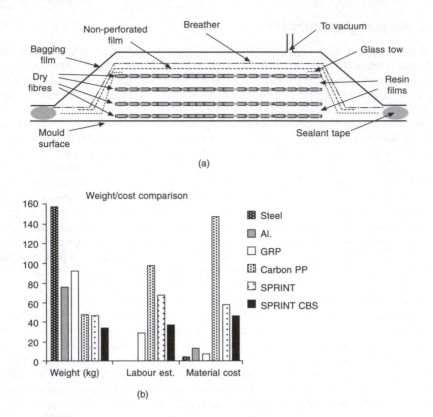

(a)

(b)

Fig. 3.17(a) and (b) Nature of SPRINT material and comparative costs. (Courtesy of SP Systems)

Fig. 3.18 Ultima body and component structure

parameters relevant to the production of body components. In Chapter 5 we will understand why these parameters are important and how they are used to optimum advantage in component production.

3.6 Learning points from Chapter 3

1. When selecting the materials for car body construction other factors such as environmental acceptability and ease of manufacture ('process chain effects') must now be considered alongside cost and the physical and mechanical properties traditionally used for engineering design.

2. Existing investment in automotive facilities and familiar design and manufacturing procedures associated with sheet steel will favour its continuing use as the predominant high volume material, and high strength and zinc coated variants will allow medium-term lightweight and durability targets to be met.

3. Downweighting the body structure beyond 30 per cent will increasingly call for alternative materials but these will require more radical changes to both manufacturing and disposal (recycling) procedures.

4. From evidence to date aluminium is the most likely contender (demonstrated by the Audi A8/A2 spaceframe architecture) to replace steel, and recycling procedures are in place to absorb most scrap and ensure that up to 50 per cent of the original cost can be recovered. Improved pricing stability regarding the initial cost of aluminium strip is required.

5. Inevitably hybrid structures will increasingly find favour with mixed material application, e.g. steel/aluminium substructures with polymer skin panels, incorporation of front end panel and hardware parts, but selection and design should allow for ease of identification of component materials plus easy disassembly and recyclability.

6. More significant weight savings may be required to boost the performance of cars propelled by electricity or alternative fuel systems and ultralightweight construction in sandwich or honeycombe forms utilizing aluminium and composite formats will be favoured. These will impose even greater constraints on process chain operations and recyclability than aluminium or composites.

7. As well as meeting the criteria used in the material selection summary Table 3.3, product consistency is essential. Maximum productivity relies heavily on uniformity of properties, dimensions and finish.

8. Coincident development of continuous strip processes for casting, annealing and finishing with associated statistical process control at each stage of manufacture must help achieve increased product uniformity.

9. Process control during forming and painting should improve with the increasing number of deterministic surface finishes being promoted by European steel and aluminium producers (EDT, EBT, etc.). The uptake of this technology depends, however, on the market availability of such finishes and standards that allow precise definition of the required topography.

10. Polymer panels have the undoubted advantages of cost (especially at lower volumes) and shape versatility but recycling remains a major problem. More effort is required in rationalizing the number of materials with selection favouring the more easily reused/recycled material type.

11. Advanced composite manufacture has been regarded as labour intensive with high facility costs but 'one hit' preprepared resin/reinforcement materials appear to offer a more straightforward route allowing advantages for higher volume sports car production.

References

1. Dieffenbach, J.R., 'Not the Delorean Revisited: An Assessment of the Stainless Steel Body-in-White', SAE Paper 1999-01-3239, 1999.
2. Davies, G. and Easterlow, R., 'Automotive Design and Materials Selection', *Metals and Materials*, Jan. 1985, pp. 20–25.
3. Sollac, *Book of Steel*, Lavoisier Publishing, Paris, 1996.
4. Price, W.O.W. *et al.*, 'Conventional Strand Annealing', Special Report No. 79, The Iron and Steel Institute, London, 1963, pp. 71–81.
5. Petch, N.J., *JISI*, 174, 1953, pp. 25–28.
6. Kennett, S.J. and Owen, W.S., 'Some Metallurgical Aspects of the Annealing of Mild Steel Strip and Sheet', Special Report No. 79, The Iron and Steel Institute, London, 1963, pp. 1–9.
7. Butler, R.D. and Wilson, D.V., *JISI*, 201, 1963, pp.16–33.
8. Butler, R.D. and Pope, R., '*Sheet Metal Industries*', Sept. 1967, pp. 579–597.
9. Davies, G.M. and Moore, G.G.,'Trends in the Design and Manufacture of Automotive Structures', *Sheet Metal Industries*, Aug. 1982, pp. 623–628.
10. Pearce, R., *Sheet Metal Forming*, Kluwer Academic Publishers BV, Dordrecht.
11. Martin, P., 'McLaren F1's Composite Body', *Engineering Designer*, Nov./Dec. 1996, pp. 10–14.
12. Thomas, S.M., 'Automotive Infusion for Composites', *Materials World*, May 2001, pp. 19–21.
13. Gatenby, K.M. and Court, S.A., 'Aluminium–Magnesium Alloys for Automotive Applications – Design Considerations and Material Selection', IBEC '97 Proceedings, Stuttgart, Germany, 30 Sept. – 2 Oct. 1997, p. 139.

4 The role of demonstration, concept and competition cars

Objective: To collate the information gained from the major development programmes of the last 20 years and from these extract the general technological trends taking place. This will enable a more structured appreciation of the various emerging technologies than is evident from specific model references where the linkages and progress between innovative steps are not always evident. Major development programmes in this context also extend to concept and competition cars where innovative ideas are often evident in feedback to production car design.

Content: The relevance of broader-based new technology programmes is introduced – a reminder is given of major projects, ECV3, ASV and proving of aluminium bonded structures – the promotion of lightweight steel opportunities via ULSAB/ULSAB 40 is explained – related national/European initiatives are highlighted – feedback is considered from concept designs and ultra-performing F1 materials – links with production models are outlined.

4.1 Introduction

Reference has already been made to major development programmes in Chapter 2, where spin-off technology can be traced to broader-based development programmes. For example, the roots of many of the modern aluminium-based/hybrid designs discussed in the latter part of Chapter 2 can be found in ECV 3 and ASVT programmes, while ULSAB extends the boundaries of steel utilization by promoting the use of an increasingly wide range of higher strength coated steels for which stiffness of flanges can additionally be increased by compatible laser welding along flanged joints. Thus while these designs have not appeared in volume production, the design principles and material choices used have had a significant influence on emerging model programmes. The same applies to concept cars, which although only built in small numbers, provide important feedback on aspects of styling, safety and overall acceptance of aesthetic appeal and allow presentation of new technology, including materials. Progressive modifications at successive motor events are common and again evidence of novel features is soon evident in subsequent generations of related marques. The introduction of vehicle innovations from competition cars following experience gained on the race track or rally circuit is well known in production models, as witnessed by the extension of carbon fibre technology to sports and luxury cars.

Again therefore, as in Chapter 2, even though some of the reference models appear dated, it is considered that these basic technologies can best be understood by an appreciation of their evolution.

The purpose of this chapter is to summarize the progress that has been made within each of these initiatives, extract the essential learning, illustrate technological links with examples that were presented in Chapter 2 and hopefully provide a reference for future application and further development.

4.2 The BL Energy Conservation Vehicle (ECV 3) and aluminium structured vehicle technology (ASVT)

Although not unique, and it is probable that other companies were seriously assessing aluminium-based lightweight structures at the same time, the BL Technology ECV 3 was a significant development programme exploring the feasibility of ultra-fuel efficient vehicles. Moreover the lessons learnt from this and similar subsequent technology are extremely instructive in understanding the material selection and processing of today's vehicles.

The Rover company had long been users of aluminium (being virtually the only material available in 1948 for the bodywork of the Land Rover) and it was natural that experience gained with this material made it a strong contender for a fuel efficient concept car. The motivation for the design stemmed from two oil price shocks in 1973 and 1979, and the ensuing uncertainty of supply and costs of oil in the future prompted BL Technology to embark on an Energy Conservation Vehicle Programme in the late 1970s. The ECV 3 was a totally new design but incorporated many of the ideas and processes from its predecessors, ECV1 and 2, and was first announced in 1982. A paper written on the car in 1985[1] commented that the concept stood up to examination after 3 years and it should be remembered that the radically different technology introduced then has been adopted by a number of production vehicles referred to in the preceding chapter.

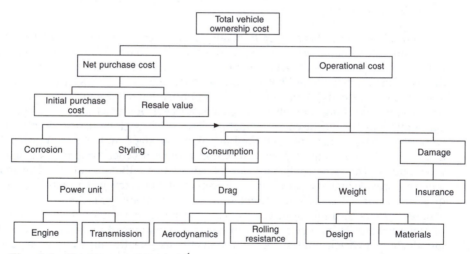

Fig. 4.1 Total ownership costs[1]

As for current new designs (see Chapter 8) the vehicle was planned with due regard to the total energy consumed in a vehicle life cycle and also total vehicle ownership costs, which were related to the factors shown in Fig. 4.1.

Apart from fuel consumption and servicing costs two other factors were identified as being of special relevance to potential owners, these being corrosion and low speed accident damage – which affects insurance costs. The body materials adopted were key to this situation and because of the relevance to current developments a short summary of the reasons for the choices made follows.

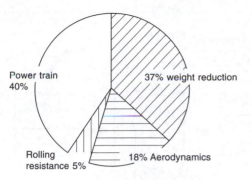

Fig. 4.2 Contributions to improved performance[1]

The concept of a clad substructure was not new and was similar to the base unit used on the Rover 2000 between 1963 and 1975. Apart from the experience the Rover company had gained with aluminium, steel was considered too heavy to achieve the objectives outlined above. Even with the most efficient use of high strength grades, now starting to emerge in the 1970s, the elastic modulus remained unchanged so opportunities for downgauging were limited to strength related parts.

Table 4.1 Choice of materials for ECV 3 and adhesive strength using various pretreatments[1]

Material / Property	Body Steel	Aluminium	High Strength Steel	Plastics		
				Polycarbonate	PU-RRIM	Sheet moulding compound
Weight/area for equal panel stiffness	1	0.5	1	0.7	1	0.6
Cost/area for equal panel stiffness	1	3	1.4	5	5	2.3
Weight for equal tensile stiffness	1	1	1	15	20	4
Energy absorbed per unit weight at break in tension	1	1.4	1.2	3.6	2.3	0.02
Strength per unit weight	1	2	2.3	1	0.5	1
Corrosion resistance	X	√	X	√	√	√

Stiffness of a structure can be greatly improved by the use of a structural adhesive rather than spot welding and it was proved in tests that torsional stiffness levels approaching those of spot welded steel could be produced at half the weight with 5xxx alloy sheet. However, to ensure durability under service conditions it is necessary to pretreat the aluminium sheet and the selection of a suitable formulation followed only after extensive accelerated tests. To guard against peel failures in impact it was necessary to use toughened epoxy adhesives plus some spot welds in flanges, although the frequency could be reduced by two-thirds compared with normal steel assemblies. At the outset it was intended to ensure a system suitable for high volume production and the process envisaged for treatment of the base unit is shown in Fig. 4.3.

For the skin panels the criteria shown in Fig. 4.4 were used.

The following materials were selected for the skin panels:

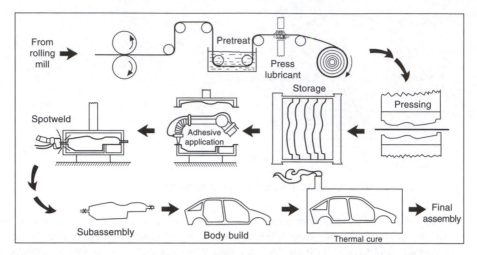

Fig. 4.3 Proposed system for the pretreatment and processing of aluminium sheet[1]

Front and rear ends	RRIM PU
Vertical skins	RRIM PU
Bonnet and tailgate	SMC
Roof	Part of aluminium structure

The advantage of the reinforced reaction injection moulding (RRIM) polyurethane (PU) is that it is a flexible 'friendly' material and resistant to minor damage, e.g. scuffing of gateposts, while the SMC has more stiffness and maintains horizontal panel contours.

The bodyweight of the ECV 3 was 138 kg compared with 247 kg for an equivalent steel struture and the vehicle weighed 664 kg or the same as an Austin-Rover Mini, yet the internal space was the same as an average mid-range European car. This weight reduction assisted in achieving all aspects of the specification.

The technology was then carried forward by Alcan who had been close collaborators on the programme, with the objective of building replicas of production cars and developing adhesive bonding technology for use at all volume levels. The first venture was to build six Austin-Rover Metros using production facilities to prove that the technologies could be applied under typical mass production conditions, and using precoated coil of the 5251-0 alloy it was demonstrated that the material could be formed using production tooling although some adjustment and rework was required to achieve some of the shapes obtained in steel. Weldbonding was carried out manually and a scheme developed that would allow robotic application in larger numbers, meeting production rates and overcoming hygiene problems with human contact with epoxy formulations. It was claimed the assembled body-in-white aluminium alloy base unit structure built with ASV technology required no further finishing to achieve a durable life[4] and painting was limited to cosmetic areas, which today would be covered by plastic mouldings anyway.

The manufacturing feasibility was thereby proven and a fuller description of the experience and modifications necessary was given in a subsequent paper by Selwood et al.[3] and Kewley et al.[4]

Further replica exercises were then carried out including five Bertone-built Fiat X1/

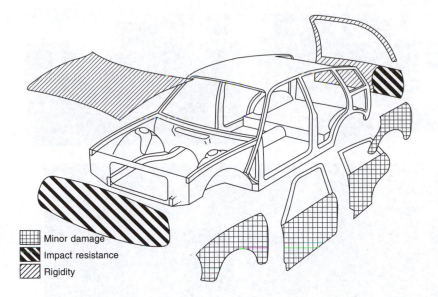

Minor damage
Impact resistance
Rigidity

Skin panel material selection					
Material Property	Steel	Aluminium	PUR-RRIM	Sheet moulding compound	Polycarbonate
Useable thickness, mm	0.8	1.0	2.5	2.5	2.5
Damage resistance (elongation at yield) %	0.15	0.2	10	0.2	6
Weight/area	1	0.45	0.5	0.75	0.5
Cost/area	1	2.4	1.7	2.2	2.6
Corrosion resistance	X	✓	✓	✓	✓

Body skin panel materials					
	Front and rear	Bonnet and boot	Sides	Roof	Limitations
Reaction injection moulding (polyurethane)	✓	✗	✓	✗	Modulus. Strength. Paint temperature Needs post mould support.
Injection moulding (thermoplastic)	✓	✗	✓	✗	Expansion Material cost? High investment cost. No way of doing valid prototypes.
Compression moulding (polyester)	✗	✓	✗	✗	Needs sealing. Brittle.

Fig 4.4   Selection of materials for skin panels[1]

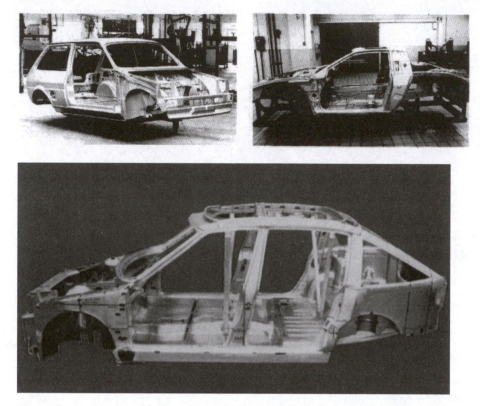

Fig. 4.5 Metro, Pontiac Fiero and ECV 3 main body structures produced with ASV technology[6]

9s, two Pontiac Fieros, the MG EX-E concept car, the Jaguar XJ 220 and the Ferrari 408. The first prototype of the latter already used a laser welded stainless steel tub with subframes bolted front and rear to support the engine, suspension and drive train. Body panels were GRP-clad polyurethane foam mouldings. Alcan introduced modifications, closing off the bottom of the centre tunnel and including doublers at points of high stress, but were able to prove that a weight reduction of 27 per cent was possible accompanied by a 22 per cent increase in torsional stiffness using aluminium bonded construction. The model, although never produced, became a rolling laboratory to evaluate future Ferrari technology.

During the replica programme described above improvements in the application of this technology were progressively introduced, including the choice of heat treatable alloy sheet (6xxx) series for more dent resistant outer panels (see Chapter 3), optimization of the pretreatment/prelubricant system applied to the sheet surface, and at a later stage, the application of riv-bonding to replace spot welds. This ASV technology has now been adopted for the latest Jaguar XJ Series each body using 3180 rivets in conjunction with adhesive bonding applied over Alcan pre-treated sheet.

Parallel developments were taking place with 'Audi Space Frame' (ASF) design and these were fully described in Chapter 3. It is appropriate to mention the embryonic phase of this technology under the concept heading; however, prior to the Audi A8 and then the A2 prototype, Audi 100s were thoroughly (and discreetly) built and tested in aluminium to determine the actual capabilities of a completely aluminium bodied car. This concept proceeded through five iterations before the A8 body structure

Fig. 4.6 ASV Ferrari 408 under construction and finished Pontiac Fiero.[6]

emerged and as previously described this was a carefully planned blend of pressure castings, extrusions and sheet either welded, adhesively bonded, riveted or clinched together. Thus an alternative aluminium bodied approach emerged and it was claimed that the scrap rate during manufacture was reduced to 15 per cent (compared to 40 per cent for a pressed steel equivalent).

As apparent in Chapter 3, this technology has advanced to selected production models including recent Audi and Jaguar models, and related technology on aluminium continues to progress following initiatives such as the BRITE/EURAM Low Weight Vehicle BE 5652 programme.

4.3 ULSAB and ULSAB 40

At the time of increasing worldwide interest in aluminium and the emergence of aluminium bodied cars such as the Audi A8, the ULSAB ULtralight Steel Auto Body initiative was launched to re-emphasize the versatility of steel and introduce new ideas to further enhance weight savings that could be achieved, not only with sheet but other product forms as well. This was a well-resourced programme supported by 35 steel companies from 18 countries who enrolled the automotive expertise of Porsche Engineering Services to prompt engineering initiatives and supervise the various validation programmes. The main findings of the programme have been described many times[7,8] and the associated follow-up programmes are still in progress. Subsequent initiatives from the ULSAB consortium have included the ULSAC project, a study of lightweight steel door and closure designs and the recent ULSAB-AVC or advanced vehicle concept project. Steelmakers believe all of these projects will produce solutions to lightweight autobody design using steel. The initial programme was reported in mid-1998 and findings are summarized in Figs 4.7 to 4.9.

The concept phase involved benchmarking against nine of the world's most popular mid-range cars and establishing key design requirements before proceeding to improve all relevant parameters to significant, yet realistic levels (Fig. 4.7).

Performance

Structural

	Benchmark	Reference	ULSAB
Static torsional rigidity (Nm/deg)	11 531	13 000	20 800
Static bending rigidity (N/mm)	11 902	12 200	18 100
1st body structure mode (Hz)	38	40	50
Mass (kg)	271	250	203

Crash analysis

ULSAB meets standards for:

- 35 mph NCAP 0° frontal
- 55 km/h 50% AMS frontal offset
- 35 mph rear moving barrier
- 50 km/h European side impact
- Roof crush

Fig. 4.7 ULSAB benchmark, expectation and performance levels[8]

By use of high strength steels, hydroformed sections and sheet hydroforming, tailor welded blanks and adopting alternative assembly methods including laser welding, it was shown that for volume production of 100 000 units a year, bodyweight could be reduced by 25 per cent to 203 kg, static torsional rigidity improved by 80 per cent, static bending by 52 per cent and first body mode increased by 58 per cent. The part count was reduced from 200 to 94 stampings and a total of 158 parts. As shown in Fig. 4.7 all five safety standards were met.

The materials and processes used are shown in Fig. 4.8.

The final assembly and achievements are summarized in Fig. 4.9.

As many automotive companies were urgently seeking weight savings in the 1990s

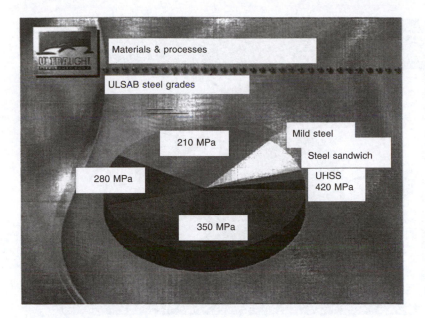

Fig. 4.8 Materials strength levels and tailor welded blanks used in ULSAB
demonstration body

the natural consequence was that much of the ULSAB type technology was being
progressively implemented anyway as a result of increasing co-operation and regular
steel development meetings between steel suppliers and design engineers. New grades
were developed in the UK and Europe that designers could utilize to achieve weight
savings and crash performance, so even in midstream the spin-off from broader
initiatives such as ULSAB were evident.[9] The original ULSAB initiative has progressed
through further phases on various autobody programmes, the latest being ULSAB-
AVC, the results of which were published in early 2002 and demonstrate even
greater advances for mid-size European and American model designs. These embrace
a wider range of 'enabling' high strength steels and conclusions are presented in
Chapter 9.

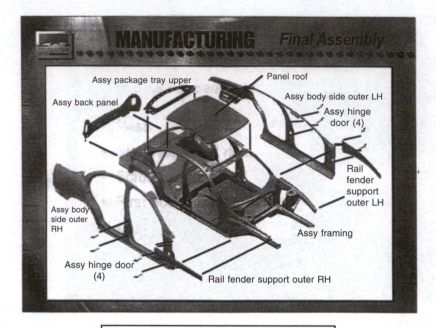

Fig. 4.9 Final assembly and materials/processes used[8]

4.4 Concept cars

The general perception of concept cars is perhaps that they are notional figments of stylists' imagination and have no real purpose beyond attracting the attention of the public to the manufacturer concerned. Sometimes these take the form of a 2-D illustration used for some of the more imaginative forms, but more often they are found as a major showpiece within the cluster of standard models comprising exhibits at national motor shows – especially when a manufacturer has no startlingly new models in that particular year!

However eye catching the presentation, the purpose is quite intentional. As well as demonstrating how 'switched on' an organization is to the latest technology, the idea is generally to attract feedback from the buying public on a particular aspect being promoted.

Following on from the ECV 3 theme, the MG EX-E was a styling exercise to maintain the interest of the public in the MG brand and was a launching pad for the ASVT technology described above, which was then followed by the Jaguar XJ-220 – both these models paving the way for public acceptance of bonded aluminium structures. Often the timing of motor shows coincides with technical conventions and exhibits linked with technical presentations, as happened with the ASVT paper referred to in Fig. 4.10 where the technology on display at the Detroit Motor Show can be explained in more detail at the accompanying SAE Automotive convention.

The chief Volvo designer considers that 'concept cars are an excellent way of providing a glimpse of the future without being constrained by a specific design. They help us make wise decisions in our development work.' Taking as an example the 2001 Detroit Motor Show, Volvo used the occasion to demonstrate its Safety Concept Car and as can be seen in Fig. 4.11 this was designed to give the driver increased control and visibility. As well as see-through 'A' posts, cutaway posts rendered partly transparent through the use of a steel box construction combined with see-through Plexiglass giving improvement in all round light ingress, 'B' posts curve inwards at the top to give the driver an unobstructed field of vision to the offset rear. In terms of safety, these 'B' posts are at least as safe as conventional 'B' posts in a rollover or side-impact scenario since they are integrated with the front seat frames. Other features include sensors that scan the precise position of the driver's eyes and adjust the seat to allow the best possible vision, and sensors embedded in the door mirrors and rear bumper that alert the driver of the approaching traffic in the blind spot to the offset rear. The headlamps monitor the car's road speed and steering wheel movements and adjust the lighting to suit progress.

The car continues to emphasize the company's awareness of occupant safety. The chief designer adds: 'By tradition Volvo was an engineering-driven company. In the past concept cars were primarily regarded as a way of presenting new technology. However, as the automotive world and the media that cover it are visually focused, a concept car also needs an innovative design if it is to attract the right attention.'

Recent exhibits at the Detroit Motor Show include the GM Precept FCEV and the GM AUTOnomy both of which point the way ahead for fuel cell technology and innovative use of materials.

The Dodge Powerbox is a hybrid concept claimed to be 60 per cent more fuel efficient than a comparable gasoline V8 version.

In Europe the Smart Micro Compact Car demonstrates a fairly radical approach to design which has progressed from the concept stage to production. As shown in Fig. 4.12 the rigidly designed Tridiron steel safety cage acts as the base unit to which nine 'easy exchange' polycarbonate body panels are mechanically fastened. Panel replacement at a Smart centre is claimed within 1 hour and the coloured surfaces resist minor scratching and offer 'optimum recycling possibilities'. Using a choice of 45–55 bhp 3-cylinder Mercedes Benz turbo engines consumption figures of 60 mpg can be achieved for unleaded petrol and 80+ mpg with the 41 bhp diesel version.

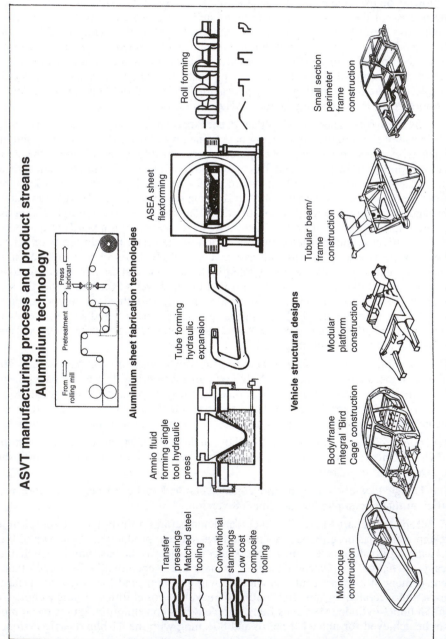

Fig. 4.10 Systems for building automotive structures using various forming methods

Fig. 4.11 Main features of Volvo's 2001 concept car

Far from providing simply aesthetic styling exercises therefore, concept vehicles also provide tangible evidence of progress being made to improve safety, and answer environmental challenges. As can be seen, aerodynamic styling is also optimized to improve the drag coefficients and in time, once benefits have been confirmed and assuming the feedback is favourable, such innovations will feature in future production, as is evident in Chapter 3.

4.5 Competition cars

4.5.1 Background to Formula 1 materials technology

The exceptional conditions under which competition cars operate provide an excellent testing ground for newer materials where performance is critical and cost is largely irrelevant. Durability, structural integrity and impact resistance are all of vital importance, the latter being dealt with at length in Chapter 8 for Formula 1 vehicles where the

Fig. 4.12 Smart Car profile

blending of composition, shape and contour of carbon fibre composites used in the construction of the main chassis is also a subject of fine tuning. All F1 materials must be ultra-low weight as every kilogram over the minimum permissible weight adds about 0.03 second to each lap so 5 kg would add 12 seconds to the total race time on a typical circuit.

The actual form of the body panel material is essentially a sandwich of aluminium honeycomb between layers of multi-ply carbon fibre reinforced resin. As well as being of composite overall construction the carbon fibre layers can also be termed composite as it is a blend of carbon fibre and preimpregnated epoxy resin. A typical chassis would comprise five major panels all bonded together. The inner and outer skins typically consist of five or seven layers of various material thickness and type, dependent on the specific strength required at different locations, and optimized by finite element analysis. A typical chassis weighs 35 kg and is capable of transferring about 750 hp to the racetrack and withstanding two tonnes of aerodynamic downforce.[13]

Formula 1 body technology is a matter for expert appraisal and for this reason the opportunity has been taken to reproduce large extracts from the latest of several accounts compiled by Brian O'Rourke of Williams GP Racing[14] (sections 4.5.2 to 4.5.26) based on many years with composites development for competition cars.

4.5.2 Introduction

The invention of the automobile gave birth to a new type of sporting activity – motor racing. During the earliest days individuals who owned cars would race them for their own entertainment. Those who were most serious about winning races realized very quickly that it was the *combination* of driver *and car* which produced results and so looked to science in order to further their ambition.

It became known that some makes of vehicle were more suitable for this activity than others. Manufacturers of those cars which were successful were quick to exploit their racing success for marketing purposes, a process that has continued to this day. At the same time it was understood that racing was an ideal activity for developing the performance and durability of a car's components. In short, motor racing has always been a target for the application and development of state-of-the-art technology and a breeding ground for new engineering ideas. Into this arena in the last three decades of the twentieth century have come advanced composite materials.

The following sections will attempt to describe the current uses of composites in the design and construction of Formula 1 (F1) cars through the experiences of Williams Grand Prix Engineering Ltd, a constructor and competitor in World Championship motor racing.

4.5.3 F1 car structures — why composites?

While composite materials in various, simpler forms had been used on motor vehicles – and undoubtedly, racing versions too – in earlier examples, what appeared in the early 1980s was the introduction of advanced composite materials to primary load-bearing applications. Most significant among these was the semi-monocoque chassis component, a tubular shell structure forming, additionally, the driver's 'survival cell'. Why was this done?

Racing cars differ from those driven on the road in many ways and one of them is adaptability. Since a car driven quickly is always at the limit of its adhesion envelope when cornering, fine-tuning of the many aerodynamic and suspension variables that are provided is an absolute necessity. The ability to make small alterations which will effect subtle changes to tyre contact force balances is highly desirable if optimum performance around a circuit is to be realized. The behaviour of the structure while it is under load is crucial to the attainment of this goal since it forms the link between the front and rear suspension systems. It has been understood for a very long time now that, in order to obtain the required sensitivity in its set-up, a racing car chassis must possess at least certain levels of structural stiffness. Simultaneously, designers have always appreciated the advantage of having a low vehicle mass, particularly those types competing on 'road' circuits (i.e. those which consist of straights and corners with varying track widths, smoothness, contour, and geographical situation) where inertia plays a key role. Also, anyone who has ever witnessed a motor race will understand that the car needs another quality – its structure must be *strong*.

4.5.4 History

The search for these requirements – stiffness, strength, minimum mass – has seen a progression of different technologies being employed in racing car construction during their evolution. Not surprisingly, the similarity to aircraft methods reflects these common objectives although their introduction to racing cars lagged behind by several, and in some cases many, years. Initially, tubular spaceframes were used until replaced by folded and riveted shell structures, a typical early example being that employed on the rear-engined Lotus 25 in 1962.[16] Following this, bonded aluminium-skinned honeycomb panels replaced the fabricated ones as typified by the Wolf WR1 F1 car of 1977.[15] It was a logical progression, therefore, that composite materials should replace the aluminium used for the face sheets in this type of structure.

There is some debate as to what the first example of a primary load-bearing composite structure on an F1 car actually was. Certainly, a rear wing support fabricated from carbon fibre and epoxy resin was used in 1975 but its failure and the accident that resulted provoked calls for caution. It is generally accepted that the first racing car chassis making use of the material was built during the period 1980–1981 and was the McLaren MP4/1.[16] It was during that period that the inadequacies of the current structures became most acute since the full 'ground-effect' underwing cars of the time (with their highly stiff suspension installations) exposed deficiencies in monocoque torsional rigidity; it became the limiter on performance improvement. The success of that first example ensured that, thereafter, carbon/epoxy composite became the only logical choice for the future and progress from then on was rapid. The last F1

Fig. 4.13 General photograph of a Williams F1 car

monocoque making substantial use of metal was superseded in 1985. At the present time carbon/epoxy composites are the materials of choice for every major category of motor racing worldwide.

4.5.5 Extent of use

The F1 car of today makes use of composite materials across an extensive range of components. Viewed from the outside, every part on display is formed from a composite material with the exception of the wheels, tyres (which, strictly, are), and braking system components. The design criteria across the assembly of components vary with their duty from, at one extreme, low mass/moderate stiffness panels forming the bodywork through crushable energy absorbers to, at the other, the maximum strength/ high stiffness primary structure that makes up the survival cell. This latter was the driver for the original use of advanced composites and it is necessary to understand its function in order to appreciate why the motor racing industry was interested in them.

4.5.6 Duty – the survival cell structure

An example of the current breed of F1 car is shown in Fig. 4.13. In order to understand how the chassis structure works it is best to gain an appreciation of the parts that make up the car assembly.

A central component accommodates the driver, the fuel tank, and the front suspension elements. It is a semimonocoque shell structure and is referred to, variously, as the 'chassis', the 'tub', or the 'monocoque', although it bears a closer resemblance to an aircraft fuselage than anything that most people would associate with a vehicle. The engine is joined to the back of this unit usually by four studs or bolts and the structure is completed by attachment of the gearbox casing to the rear face of the engine. The chassis, engine, and gearbox, therefore, form a 'box-beam' structure which carries the inertial loads to their reaction points at the four corners of the car as depicted in Fig. 4.14. Arranged around, and attached to, these are the remaining components –

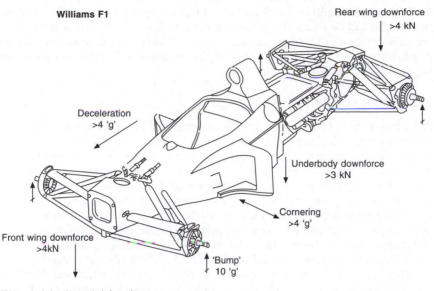

Fig. 4.14   Inertial loading

wing structures, underbodies, cooler ducting, and bodywork, as illustrated in Fig. 4.15. This general arrangement is exactly that as has been used by most single-seater racing cars since the 1960s.

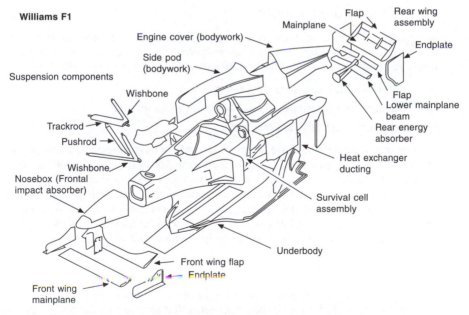

Fig. 4.15   Extent of use of composites – exploded assembly of car

It has been mentioned that the chassis component is of major importance to the working of the structure. During the course of 'setting up' a racing car at a circuit, changes are made to the suspension elements (springs, dampers, antiroll bars) with the intention of modifying its handling. Ideally, any small change in a component stiffness should be felt in the balance of the car. This will not occur if the structure transmitting the loads is of insufficient stiffness. The chassis member must, therefore, possess good stiffness characteristics or the handling will suffer and speed around the circuit will be lost.

4.5.7 Rule conformity and weight

As in any other field of engineering, the designers of F1 cars must comply with a set of regulations when arriving at their solutions. These are defined by the FIA, the governing body for motor sport worldwide. The key parameters covered by them for the structural designers are those of geometry, strength performance, and weight. Constraints are placed on the overall dimensions of the car's bodywork (which includes wings) and the sizing of the driver envelope within the cockpit. Load cases are specified for the design of key elements of the structure and tests are defined which must be performed and passed in the presence of an appointed witness. Of major significance also is the regulation which limits the minimum weight of the car to 605 kg, inclusive of driver. To achieve this while fulfilling all of its necessary functions is difficult but vital if the car's speed is to be maximized and, therefore, all of the components, including the structure, must be of minimum mass. It is reckoned that 20 kg of excess weight is equivalent to a time difference of 0.5 s around a typical circuit – several grid positions in a qualifying session or half a lap over the course of a race. These parameters, along with the changes in allowable engine specification, have been modified regularly over the years in order to contain the pace of development in car performance and improve driver safety.

4.5.8 Structural efficiency

Summarizing, the chassis must have good stiffness, be of sufficient strength to satisfy the loading requirements, be of a damage-tolerant construction, and have minimum mass. In short, the structure must be efficient. This may be maximized by:

(i)   optimizing the structure's geometry,

(ii)   selecting the most effective construction method, and

(iii) using the most efficient materials.

It is this search for maximum structural efficiency that has brought about the progression of different technologies seen in racing car construction during their history as described earlier. The attraction of carbon/epoxy composites was their efficiency relative to aluminium in that they possess better specific moduli and their properties may be tailored to the needs of the structure. Components are produced by a moulding process which allows the full external geometry to be used as a working structure (i.e. there is no need for separate, covering bodywork) and so furthering effectiveness.

4.5.9 F1 – A good match for composites

It is often overlooked by people working in the F1 industry that they are, actually, fortunate in terms of how well the composites that they use fit the profile of what is required. The materials chosen are supplied in prepreg form and the processes used are vacuum-bag and autoclave curing followed by machining, manual trimming, assembly, and bonding. The production route, therefore, is a very labour-intensive one which, coupled with the high initial cost of the materials, makes for very expensive components. It also dictates that parts can only, sensibly, be made in relatively small quantities. This matches very conveniently the often-changing, but rarely high unit-number, nature of F1 car component production. Again, if the industry were not able to tolerate high unit cost parts this combination of materials and processes would be judged wholly inappropriate. In short, it is only the high-value/small-volume region of the cost envelope in which Formula 1 car production lies that allows composites of this type to be effectively employed. The parallel with military aircraft structures – also heavily reliant on composites – here is an obvious one.

4.5.10 Design

The design of an F1 car makes extensive use of computer-aided engineering. This covers aerodynamic design, geometry definition, drawing production, structural and fluid dynamics analysis, and master pattern machining. The production mould tools are taken directly from these patterns.

The design solution for the F1 car monocoque is illustrated in Fig. 4.16. It consists of five principal components. The outer shell of the structure is reduced to two; a separate, largely flat, floor panel being joined to the remainder at its base. Bulkheads are positioned so as to feed suspension point loads into the structure and enclose the cockpit bay. Attachments fit into solid inserts bonded within the shell honeycomb. It forms part of the aerodynamic envelope of the car and so is, of necessity, often a complex shape.

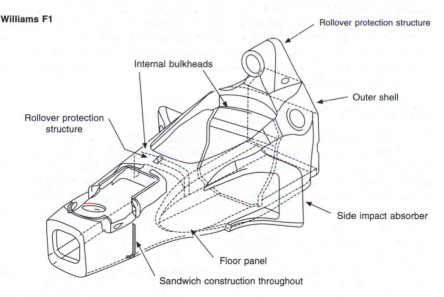

Williams F1

Fig. 4.16 Monocoque assembly

Since one of its primary functions is that of forming the driver's 'survival cell', when considering the design of the chassis structure a balance must be found between the goals of weight, stiffness, and strength. Today's circumstances are such that there is a real premium to be had from minimizing weight as much as possible, even to levels which will give a car considerably below the allowable limit since ballast may then be added in locations where it is most beneficial for handling. In terms of the author's own philosophy, the process should be: (i) determine, as part of the specification, what are the minimum levels of stiffness required for adequate handling characteristics; (ii) produce a design solution which satisfies these and, for that necessary mass of material, include the best combination of material properties, geometry, and manufacturing details to provide the maximum possible strength solution. The point of the exercise is not merely to satisfy the mandatory strength demonstration cases but to design for the real ones in which loading cannot be so easily quantified. A rudimentary fact is that simply seeking to maximize stiffness brings limited returns above a certain level; beyond this attention should be concentrated on strength.

Wing structures, in a similar manner to the survival cell discussed, must embody damage-tolerant thinking since they are also safety-critical components. While aerodynamic variations due to deflection must be minimized, strength considerations should not be neglected when choosing materials or a manufacturing route.

The more recent extension of composites for use in selected suspension elements has provided the first examples of components which are strength-designed as a primary requirement. The choice of an appropriate manufacturing method in combination with careful analysis of the structural function involved has demanded considerable effort to evolve a suitable design solution. It has to be stated, however, that fully composite suspension systems do bring questionable benefits in impact situations relative to the traditional tubular steel ones that they have replaced. In contrast to the composite energy absorbers discussed later – where controlled in-plane brittle fracture is a positive feature – a 'spaceframe' structure such as constitutes a suspension assembly can only give rise to out-of-plane element deflections during an impact. In these circusmstances, the ductility of metal tubes giving rise to plastic 'hinging' after buckling allows plenty of scope for energy absorption with which the single fracture that occurs for composite equivalents cannot compete. It is also impractical to expect to match bending stiffnesses with the lower modulus of carbon/epoxy materials and this dictates a choice being made between accepting a degradation in performance or increasing section geometry with its attendant aerodynamic implications in order to realize a weight saving.

4.5.11 Chassis loading

The chassis structure is subject to severe inertial loading. Currently, a fully laden Grand Prix car may be subjected to sustained loads of 4.5 g laterally, at least 4 g under braking, and as much as 10 g as instantaneous 'bump' loading at one or more of the wheels. The nature of these actions is constantly varying and upon which cyclic high-frequency inputs from sources such as engine vibrations must also be superimposed. Aerodynamic down-force from the front wing is input to the structure at the nose attachment points. This may be as much as 4 kN. Additionally, the chassis must be designed to cope with the impact test and strength demonstration cases. All of these loads are summarized in Figs 4.14, 4.17 and 4.18. The structure that results from the criteria detailed here weighs approximately half that of the driver that it is built to accommodate.

4.5.12 Analysis

The descriptions of the components and processes presented serve to illustrate that the functions and construction of the types of structure used in F1 motor racing do not lend themselves easily to simple analytical cases. While finite element analysis may be employed in panel design and 'failure indices' based on a chosen criterion obtained, the current state of practice is to regard these results, although valuable, as being only part of the process. The demonstration of the suitability of a design is still determined by the mechanical testing of key structural elements and, wherever possible, a complete structure as proof of its integrity. Real data measured while a car is

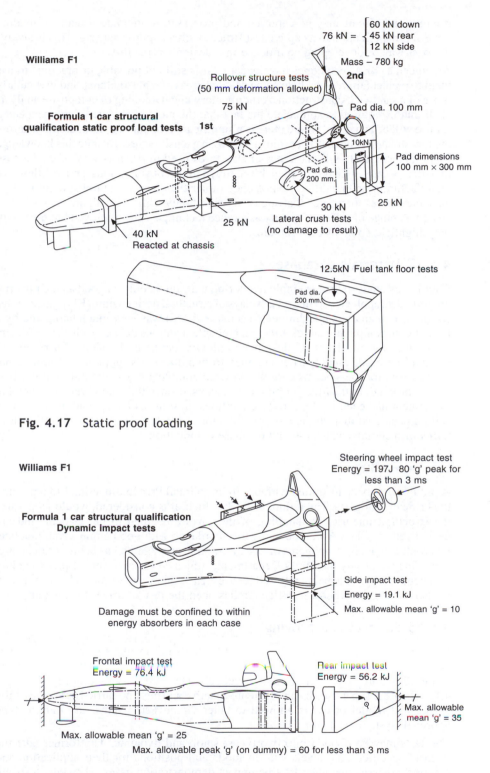

Fig. 4.17   Static proof loading

Fig. 4.18   Impact proof loading

running at a circuit may be captured and processed to emulate a realistic loading system for application to a captive test structure on an appropriate rig. This is always the best way of developing confidence in a design's capability.

Advanced as FEA techniques have become, it is still not possible or practical to use them to predict failure quickly in some detailed areas of real structures, and particularly those of a sandwich construction. A rudimentary understanding of composite analysis is all that is required to appreciate the fact that the most likely site of an unexpected failure will be at a detail where there is a significant amount of loading applied across the low strength resin matrix or joint adhesive in a tensile sense. Despite this knowledge it is inevitable that, with the types of structures used in F1, it is very often the case that their shape will be dictated by aerodynamic considerations and by those of manufacturing practicality or expediency rather than best structural suitability. It is also inevitable, therefore, that areas of uncertainty will be built in to any given design. A range of different quality assurance techniques are employed but these will only highlight some of the problems.

4.5.13 Materials database

Clearly any analysis that is employed during a design is only as good as the material property data applied to it. In the early days of structural design within F1 organizations, confidence in analytical techniques (such as existed) was very much hampered by a lack of both computational resources and reliable materials data, as well as insufficient manpower for the level of design complexity being undertaken. Carrying out calculations based on properties obtained from a material's supplier – assuming that they existed for the systems chosen and were complete – was the only but not best way to proceed. In the time period that has elapsed since then, however, considerable investment has been made in testing facilities, design resources, and appropriately skilled people so that, in the case of Williams at least, we may be self-sufficient in terms of materials evaluation and database compilation.

4.5.14 Testing

A protocol has been instituted within Williams Grand Prix Engineering Ltd regarding the criticality of parts. The failure of those which, it is considered, would jeopardize the structural integrity or handling of the car are deemed to be 'critical' and are designated as 'Class A' components. In accordance with good engineering practice, it is routine that all components or assemblies that are regarded as being 'critical' are, following release by the Stress Department, subjected to some sort of proof loading or function test prior to being released for use on a running car. Stiffness characterization is also carried out wherever deflection has been the design driver for the part.

4.5.15 Survival cell proving

In addition to the procedure above, the FIA have put in place a series of 12 tests which are intended to demonstrate that a minimum level of crashworthiness has been achieved within the design of the survival cell. They are illustrated in Figs 4.17 and 4.18. These are supervised by an FIA representative and must be carried out successfully before that model of car is allowed to enter Grand Prix events; in effect a 'type approval' system exists for F1 cars.

The tests may be divided into two types: static and dynamic. The former take the form of specified load levels with methods and positions for their application and reaction, whereas the latter are true impact demonstration cases. The static tests are further subdivided into those where every example of a structure built must be loaded and the remainder where a single destructive test is performed on the datum unit, i.e. the 'reference' structure that must be declared at the start of the manufacturing sequence and used for the complete set of tests. The others built must, when checked, show stiffness values that match, within a specified tolerance, those of the reference example.

4.5.16 Survival cell crush and penetration

The strength performance of the survival cell in terms of lateral crushing is checked by the application of 'squeezing' loads to one of its sides and reacted at the other. These tests must be carried out on all monocoques built and demonstrate that, during loading, deflections are contained within a specified level and that no damage results. There are four positions designated: the first is specified as being equidistant between the front wheel 'axle line' and the forward rollover protection structure; the second is aligned with the driver's harness lapstrap anchorage; the third is at a point on the cockpit edge related to the opening template; and the last is positioned to coincide with the centroid of the fuel cell compartment. The second position calls for 200 mm diameter circular load pads and 30 kN of proof load, the third 100 mm diameter and 10 kN, while the first and fourth require rectangular pads of dimension 100×300 mm for 25 kN loads. A fifth case is specified to check the fuel tank bay against penetration from below and cells for 12.5 kN to be applied on a 200 mm diameter pad, to be reacted through the engine mounting positions.

The three remaining static tests are performed once only on the reference survival cell. The structure must contain rollover protection features in front of and behind the driver. These must be proof loaded to demonstrate their capability and 50 mm of deflection is allowable in each case, meaning that some crushing of the composite structure is acceptable if it can be tolerated. The load in the first case is 75 kN applied in a vertical direction, whereas that for the second is a vector summation of three components (down, rear, and side) totalling 76 kN. The third load is a simple lateral one of 40 kN applied to the nosebox which must not result in failure of its skins or connection to the monocoque front.

4.5.17 Survival cell impact

The survival cell must demonstrate its ability to withstand impact situations by successfully completing three types of test. The first pertains to frontal impact performance in which a fully representative chassis and nosebox must be subjected to an impact of 76.4 kJ energy. The structure is mounted on a sled, the combined mass of which must be 780 kg (equivalent to a fully laden and overweight F1 car) and be propelled into a solid, vertical barrier at 14 m s^{-1}. A simulated fuel load and a dummy driver must also be incorporated in order to check the integrity of the seat-belt anchorages and fuel tank bulkhead junctions. This energy must be absorbed by the structure and contain the damage within the nosebox itself while not exceeding an average deceleration of 25 g.

A second test is intended to demonstrate the structure's capability in the event of a side impact. Here, the monocoque is held rigidly and stationary while a moving mass is projected into it. The impactor face is of dimensions 500 mm high $\times$ 450 mm wide and is positioned with its centre at a longitudinal station related to the cockpit opening, itself defined as being of a minimum size by regulation. The impacting mass is 780 kg, the velocity is 7 m s^{-1} giving an energy of 19.6 kJ, and the mean deceleration must not exceed 10 g during the event. In reality, this test is a difficult one for the structural designer to satisfy since the figures quoted correspond to a minimum crush distance of approximately 250 mm. The shapes allowed for the energy absorber, however, mostly correspond to that of the front of the car's sidepod intake which do not represent the ideal that might be had. The diffusion of the resulting loads into the chassis structure behind the energy absorber are, similarly, a challenge to design since the predominant load direction is normal to the monocoque side panel which must be supported laterally by some kind of substructure or internal beam.

Rear impact protection is the third requirement of the technical regulations and is achieved by the provision of an energy absorber positioned behind the car's gear-casing assembly, usually doubling as a rear wing support structure. The test for this case requires that a gearbox, impact absorber, and rear wing assembly are fixed to a stationary barrier and a moving mass is projected into them. The mass of the impactor is 780 kg, the velocity is 12 m s^{-1} giving an energy of 56 kJ, and the allowable mean deceleration is 35 g. Again, for the optimum aerodynamic and mechanical layout of the car, the energy absorber component will usually be required to be minimal which presents challenges when maximizing its efficiency.

A final, fourth dynamic test case is that of the steering column which may involve composite materials in its construction. This calls for the steering wheel/column assembly to be subjected to a 200 J impact by a hemispherical object to simulate the driver's head striking it during a frontal accident. The failure criterion relates to an 80 g exceedance for more than 3 ms. This allows some scope in terms of design solutions as its location internal to the monocoque does not impose any geometric constraints resulting from aerodynamic considerations.

4.5.18 Impact absorber design

Clearly, the impact conditions described above require considerable thought and effort when undertaking the task of designing the energy absorbers. The requirement to provide such components was first introduced for the 1985 season when the use of structural composites on F1 cars was still relatively new; the added task of making them deform and absorb energy during an impact was completely so. In the time period since then, however, a considerable amount of experience has been acquired in this field and has been advanced by the gradual extension of the regulations governing testing from the original, single, 39 kJ frontal impact to the full spectrum of tests described above.

The question at the outset was – and today still is – how does one make use of 'brittle' composite materials in structures of this type to absorb the energy levels defined? It was found by constant experimentation that sandwich-stiffened carbon/epoxy composite panels could, when subjected to end-wise loading, fail in compression in a manner which was stable and progressive, so providing a controlled retardation. This would allow us to achieve a useful force–time history characteristic during the impact event and so optimize the component for minimum mass or crush distance.

There are many factors influencing this crushing process which are necessary to understand in order to achieve an effective result: skin/core thickness ratio, overall panel size, material properties, failure modes and, above all, the geometry of the component. This design knowledge has been compiled over many years of work and different designs, during most of which the learning was empirical. In recent times and after considerable effort in research, at Williams' real success has been achieved with dynamic finite element modelling of the crushing process. The quality of the correlations found between simulation and reality has been such that it is now possible to explore many laminate variables before committing resources to the moulding of test pieces.

4.5.19 Construction

The manufacturing processes chosen for F1 car construction comprise, in the main, hand lay-up and autoclaving of thermosetting prepreg composites. Smaller use has also been made of other technologies such as filament winding and compression moulding and, undoubtedly, these will feature more in the future.

4.5.20 Tooling

An undisputed fact of composites manufacturing is that the quality of a component is heavily influenced by the quality of the tooling used in its production. The majority of tooling for F1 operations is itself of a composite nature since geometry is complex, timescales are short, and total production quantities are comparatively small. Although much good work was done with wet lay-up carbon/epoxy moulds in the transitional period, today low-temperature tooling prepregs are used exclusively. This technique allied to fully CAD-defined geometry and five-axis master pattern machining allows for the rapid production of new designs. The pattern materials used are of an epoxy-based nature, chosen for optimum compatibility with the curing of the epoxy tooling matrix.

Mould design is another factor contributing to the quality and production rate of a component. A mould may produce a good part but is of little use if it requires a long time to turn around between units. Allied to this is the choice of vacuum bag used. Extensive use is made of rubber reusable bags where time may be saved or reliability improved by avoiding the frustrations of conventional nylon films. Consumable materials, otherwise, are of the types commonly used with similar composites in other industries.

4.5.21 Materials

The structural composite materials used in F1 cover a wide spectrum of commercially available fibre and matrix types. The majority of applications make use of epoxy resin and carbon fibres but examples of phenolic and bismaleimide matrices are common. Similarly, aramid and polyethylene fibres have found uses alongside carbon fibres of various types. As is to be expected, work began with the high strength (or standard modulus) types of carbon fibres such as XAS, T300, and HTA; the most common available. Today the necessity to meet the demands of impact worthiness and, generally, maximum strength requirements of the driver survival cell have provoked a move to intermediate modulus carbon reinforcements used in conjunction with the latest toughened epoxy matrices. Examples do still exist in the industry, also, of high modulus carbon fibres being used in what should be described as strength-designed structures, despite the many unanswered questions as to their suitability on the grounds of low compressive and interlaminar characteristics.

Metal matrix composites have also found uses in F1 racing cars where they are considered appropriate. They take the form, in the main however, of particulate-reinforced types and their use is viewed very much as an extension of the conventional metal machining process.

Honeycomb materials are used in almost all of the forms commercially available. The complex geometries of some of the components, particularly the chassis, have resulted in some combinations of types being chosen for purely manufacturing reasons.

Aluminium honeycomb is used in both hexagonal cell and 'Flex-core' forms, whereas Nomex material is procured in standard hexagonal and overexpanded types. The latter is most useful in areas where the curvature is in one direction only, while Flexcore will cope with regions of complex double curvature. Choices are made on the basis of convenience for manufacture rather than structural optimization, it being reasoned that panel bending stiffness is controlled in the main by skin in-plane modulus rather than core shear stiffness.

4.5.22 Quality assurance

A challenging field within F1 composites manufacturing is that of quality assurance. While much of the physical and geometric inspection may be carried out using many of the conventional, general-purpose pieces of equipment available such as co-ordinate measuring machines and fibrescopes, non-destructive evaluation (NDE) is more difficult to apply. Attempting to read across from standard paractices used in the aerospace industry was soon discovered to be inappropriate for the types of structural component met most often. The number of parts where fully water-immersed through-transmission ultrasonic examination may be possible is small. Instead, an emphasis has to be placed on portable equipment which will allow results to be obtained from complex curved surfaces accessible, often, from one side only. This is a difficult set of criteria to easily satisfy.

The types of NDE most commonly used within Williams are those based on pulse-echo ultrasonics or low-frequency techniques such as mechanical impedance, resonance, or velocimetric methods. A means of scanning, presenting, and recording results is also desirable; here use is made of the ANDSCAN system developed in the UK by DERA. Additionally, fixed-image radiography is used where the integrity of critical junctions may need verification. Real-time systems have been employed similarly.

The most fundamental aspect of any sort of quality assurance system applied to composites relates to manufacturing procedural discipline. F1 cars carry human beings in the same way that aircraft do and they deserve equal respect in terms of the quality of design and manufacturing applied to their construction. Every effort is made to ensure the integrity of a design through analytical techniques followed by proof load testing and, in some instances, fatigue life demonstration. Similarly, in their manufacturing sequence, emphasis is placed on good prototyping of a process prior to full production, for which procedures are carefully documented. Simultaneously, prepreg material life is monitored to ensure conformity with specification, operator details and cure cycle data are recorded so as to provide full traceability for the component.

4.5.23 Resources

The proportion of a current F1 car that is built from composite materials demands that an equivalent amount of a competing team's resources are allocated to their production. In order to put this into perspective, the more successful F1 constructors (not including the engine-supply partner's activities) at the time of writing employ upwards of 300 people to cover all of the aspects of car design and construction. The workforce engaged in the production of composites parts will amount to as many as 50 of these. The manufacturing facility given over to this work will include a good specification clean-room, separate areas for trimming, assembly, release-coating, bonding and, probably, three or more autoclaves supported by air-circulating ovens of varying size.

4.5.24 Timescales

The work carried out with composites by a user in the F1 industry has been described in some depth. When assessing this and the extent of its meaning, the reader must understand something of the timescale demands that exist. The nature of the activity is one of continuous technical development throughout the racing season with a step-input during the period in which the design of the following year's car takes place. Since it is important to incorporate the very latest thinking at its conception, the starting time for each programme is usually late summer. In order that sufficient time is allowed for evaluation and durability proving before the beginning of the season in March, the car must be substantially complete in mid-winter.

This translates to a maximum 4-month period in which to conceive, analyse, detail, produce, and prove the design of the complete set of parts. Despite the fact that many details and ideas are the result of a continuous evolution, it is very seldom that a component is moved from one generation of car to the next in its entirety. Every part, therefore, requires geometry, laminate, and detail definition from 'scratch' each time around. Given the number of components in the complete assembly this translates to a very large amount of master pattern and tool laminating as well as production of the parts themselves. In short, a considerable amount of work takes place in a short space of time – and everything is expected to be correct and function properly first time; no slack is built into the programme for problems. The satisfaction felt when a project is completed and results in a successful car, however, is considerable and, undoubtedly, provides the best motivation for further effort.

4.5.25 What has been achieved?

How could one best sum up the extent of the work done with composite materials in F1 to date? A wide range of component design specifications has been set out and met including criteria covering stiffness, strength, mass efficiency, environmental tolerance, fatigue loading endurance, and crashworthiness. Manufacturing solutions have evolved to include extremely complex shapes with good repeatability and consistent quality while supplying meaningful quantities of parts in short timescales.

To be properly objective one must look at the present situation in terms of what is lacking. Making comparisons is always difficult where the applications are not exactly parallel but in relating our activities to, for example, the aerospace industry it is certainly the case that the level of analytical understanding of our components is more rudimentary. The complexity of the shapes chosen means that there are real limits to the accuracy of structural performance that can be predicted given the methods that are currently available. Laminates for, say, an aircraft wing skin would expect to be fully optimized, whereas those for an F1 monocoque are far from that given the diversity of loading cases that it must support. Certainly, simple torsion, bending, and crash situations will be understood and designed for but optimization across the full range of cases – between which conflicts occur – has not been achieved at the time of writing – within this organization.

Set against this, the amount that has been learnt in the period of, at most, 19 years and the level at which composite structures are performing in F1 compares well against other industries. In some fields new uses for the materials have advanced extremely rapidly – the work on impact absorbers is a prime example of this.

In the author's opinion, the foremost contribution that F1 has made to furthering the use of composites generally is the exposure that it has given them to the general public worldwide. Global audiences for each F1 race are in excess of 400 million; they are seeing advanced composites in demanding situations every 2 weeks throughout most of the year. At the time of writing there are many ordinary people who like to possess objects which they see as being 'high-tech' and the inclusion within them of some carbon fibre composite materials is part of that whether they be cars, bicycles, or briefcases. While the author would not cite F1 cars as the main reason for this trend, it must have played a part in its evolution.

4.5.26 Future developments

In looking forward, what does the future hold for composites in the F1 industry? From the author's perspective there is a correct path of development which, based on a further and deeper understanding of the science governing their behaviour, will bring about improved component design and integrity. Extending manufacturing techniques to encompass, for example, resin transfer or resin infusion moulding is a logical path to follow where it is suitable. Likewise, better integrated, composites-tailored CAD/CAM software and analysis tools will improve efficiencies.

There is a parallel course, however, which could result in more examples of inappropriate applications and a general drift towards 'fashion' design. There are a good number of well-trained, specialist composite engineers working on F1 today and the route chosen should be that resulting from their decisions alone. They must be consistent in their principles and resist being directed by the larger number of engineering generalists (ever present) whose actions are governed often by the search for 'magic solutions' to car performance.

It is inevitable that the integrity, and appropriateness, of future composites structures used in F1 will reflect what is dictated by the technical regulations. The FIA, therefore, have a major role to play in framing them in ways such that good engineering solutions to the problems of, in particular, impact loading result. This is currently a subject of some debate since the bizarre situation exists today in which the rules governing the car's geometry conflict with the best provision of side-impact protection; successful solutions to the side-impact test have evolved which would be ludicrously ineffective in the event of a *real* occurrence. The motives for this are all related to the overall design being mostly controlled by aerodynamic considerations. Similarly, the definition of the car's minimum weight needs to be given attention if the current level of chassis strength is to be maintained; it could be argued that 'loopholes' existing with respect to testing allow teams to degrade its safety performance in the pursuit of mass distribution optimization. There have been many instances when prominent engineers from F1 have replied to questions with the view that safety is the most important criterion in a car's design. The reality is that, in private, the same people will admit that it is very much secondary to car performance; it being cited that a driver, if asked to choose, will always opt for the quicker solution than the safer one.

Clearly, what the composites engineer must do is seek to be part of a team whose intention is to build the best performing car while making no compromises over its safety and this will only be realized when the technical regulations are adjusted to allow for these two conflicting parameters to balance. The driver survival cell should only ever be considered as a maximum-strength design case and solutions where this is compromised by a search for greater stiffness or further decreased weight – and hence the trend towards stiffer but less strong materials choices – eliminated. Nowadays the professional accountability of engineers in F1 is well understood and it would only take one high-profile example of bad design to set the industry – and the public's perception of composites – back a long way. Safety should never become competitive.

In summary, the best way forward is through better understanding of composites, avoiding gimmicky applications and tighter regulation to promote truly safety-oriented structures while still allowing scope for innovative design. As was explained in section 4.5.2, the origins of motor racing involved the uses of engineering to further sport. This has evolved into an engineering industry built *around* sport and so – with the sole, emphatic, exception of safety provision – it must continue.

4.5.27 Hypercars

Only limited use of carbon fibre composites has become evident in actual production cars (see Aston Martin Vanquish, Chapter 2) but this is regularly mentioned in the guise of the Rocky Mountain Institute's 'Hypercar' design concept[17] which combines an ultra-light and ultra-low-drag platform with a hybrid-electric drive system. Computer modelling performed at the Hypercar Centre predicted that such vehicles of the same size and performance as the four to five passenger cars of the mid-1990s could achieve three times better fuel economy.

Figure 4.19 reproduced from the RMI paper[17] illustrates how the synergies between

- 63 per cent lower mass
- 55 per cent lower aerodynamic drag
- 65 per cent lower rolling resistance
- 300 per cent more efficient accessories (lighting, HVAC, audio system, etc.)
- 60 per cent efficient regenerative braking (i.e. braking energy recovered), and
- 29 per cent efficient hybrid drive

could improve a 1990 production platform's fuel economy during level in-city driving.

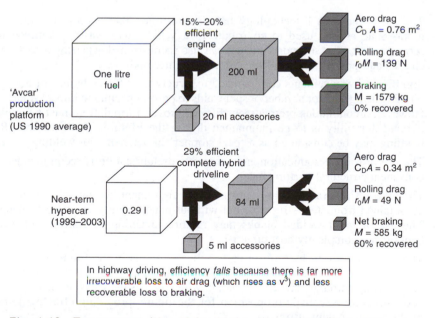

Fig. 4.19   Two ways to drive 12 km in the city – according to RMI

In addition to the above attributes hypercars would employ composites that embed reinforcing fibres (e.g. carbon, aramid, high strength glass) in a polymeric matrix. It is claimed that these have outstanding properties for autobody use including fatigue and corrosion resistance, highly tailorable material properties, generally low coefficients of thermal expansion, good attenuation of noise, vibration and harshness and precise formability into complex shapes. Materials experts from various carmakers estimate that an all-advanced-composite autobody could be 50–67 per cent lighter than a current similarly sized steel autobody as compared with a 40–55 per cent mass reduction for an aluminium autobody and a 25–30 per cent mass reduction for an optimized steel autobody.

Furthermore, the secondary weight savings result from the better performance, allowing the frugal use of those materials combined with less capital intensive manufacturing and assembly, help overcome the cost-per-kilogram premium over steel. The environmental benefits are referred to in Chapter 7.

While such an exercise is thought provoking (and many of the attributes of carbon fibre composites highlighted above have been proved in F1 competition) criticisms are inevitable. Doubts are cast as to handling under adverse weather conditions, towing performance, and the necessity for increased weight associated with complex features such as hybrid drive systems plus over optimistic modelling of braking energies. The high cost of carbon composite material is also constantly referred to, as are the recycling and end-of-life problems referred to in Chapter 8. Similar studies are already undertaken within large automotive organizations but as with ULSAB, ASVT and the other initiatives highlighted above, these external stimuli are essential to ensure designers and suppliers are aware of future possibilities and can respond with suitably modified designs and costs.

4.6 Key learning points from Chapter 4

1. The ECV and ASVT technology has demonstrated that a pressed aluminium monocoque can be used to contribute to a vehicle with vastly improved fuel economy. Important secondary savings can also accrue from downsizing of associated chassis and power train resulting from the lighter body.

2. Significant improvements can be made to structural stiffness by use of adhesive bonding using a minor number of spot welds to prevent peeling in impact situations. However, a continuous pretreatment film of proven formulation must be applied if bond durability is to be maintained during the lifetime of the vehicle. Riv-bonding may be considered as a 'peel stopper' in place of spot welding.

3. The use of robotic application and automatic prelubrication is recommended for consistent structural performance.

4. The ULSAB programme has confirmed that significant bodyweight savings can be obtained from steel structures but without major facility or process changes. The use of tailor welded blanks may require revisiting to confirm functional benefits are completely cost effective.

5. The increased use of hydroform parts appears logical in the future as structural evaluation has demonstrated gains in torsional stiffness. Tube hydroforms also offer savings through parts consolidation which makes increased utilization in the body structure a realistic proposition for the future, providing effective joining methods can be demonstrated.

6. Assumptions should not be made regarding strength levels achieved by cold working during hydroforming. Strength can vary locally around the tube circumference and there is some evidence that cyclic softening can occur. FLDs derived for sheet are not necessarily valid for hydroforming and maybe stress is a better indicator of criticality than strain (see Chapter 5).

7. Concept cars have a role in gaining acceptance by the public for future design features, the themes of which may vary from safety, weight savings to alternative propulsion methods.

8. Competition cars, demanding materials with exceptional properties, provide excellent feedback under extreme conditions, and this technology is often incorporated in future designs. More exotic materials such as carbon fibre composites, although exceptionally well proven, deserve more competitive costs!

References

1. Kewley, D., 'The BL Technology ECV 3 Energy Conservation Vehicle', SAE Paper 850103, Detroit, 1985.
2. Curtis, A., 'Techno-Triumph', *Motor*, Oct. 22, 1983.
3. Selwood, P.G. *et al.*, 'The Evaluation of an Aluminium Bonded Aluminium Structure in an Austin-Rover Metro Vehicle', SAE Paper 870149, Detroit 1987.
4. Kewley, D. *et al.*, 'Manufacturing Feasibility of Adhesively Bonded Aluminium for Volume Car Production', SAE Paper 870150, 1987.
5. Wheeler, M.J., 'Aluminium Structured Vehicle Technology – A Comprehensive Approach to Vehicle Design and Manufacturing in Aluminium', SAE Paper 870146, 1987.
6. Lees, H., 'Light Fantastic', *Autocar and Motor*, 30 Nov. 1988.
7. Ashley, C., 'Steel Body Structures', *Automotive Engineer*, Dec. 1995, pp. 28–32.
8. ULSAB UK Launch Presentation, Heritage Centre, Gaydon, 25 Mar. 1998.
9. Davies, G.M., Walia, S. and Austin, M., 'The Application of Zinc Coated Steel in Future Automotive Body Structures', Fifth Int. Conf. on Zinc Coated Steel Sheet, Birmingham 1997.
10. Walia, S. *et al.*, 'The Engineering of a Body Structure with Hydroformed Components', IBEC Paper 1999-01-3181, 1999.
11. Boyles, M.W. and Davies, G.M., 'Through Process Characterisation of Steel for Hydroformed Body Structure Components', IBEC Paper 1999-01-3205, 1999.
12. Zhao *et al.*, 'A Theoretical and Experimental Investigation of Forming Limit Strains in Sheet Metal Forming', *Int. J. Mech. Sci.*, Vol. 38, No. 12, pp. 1307–1317, 1996.
13. Macknight, N., *Technology of the F1 Car*, Hazelton Publishing, 1998.
14. O'Rourke, B., 'Formula 1 Applications of Composite Materials', *Comprehensive Composite Materials*, Elsevier Press, Oxford, pp. 382–393, 1999.
15. Noakes, K., *Build to Win. Composite Materials Technology for Cars and Motorcycles*, Osprey Publishing Ltd., London, 1988.
16. Nye, D., *History of the Grand Prix Car 1966–91*, Hazelton Publishing, Richmond, UK, 1992.
17. Fox, J.W., 'Hypercars: A Market Oriented Approach to Meeting Life Cycle Environmental Goals', SAE Publication SP-1263, Feb. 1997.

5 Component manufacture

Objective: The purpose of this chapter is to introduce the key parameters influencing material performance on conversion to the component form, and to describe the main manufacturing processes involved. The main focus concerns the primary shaping of materials but the subsequent operations are discussed where relevant.

Content: Background essentials of modern high production pressworking are introduced and parameters influencing formability are defined – the derivation of test values are explained and the significance of forming limit diagrams and use is summarized – an explanation is given of the influence of different steel surface topographies – main form and cutting tool materials are introduced together with heat treatment and repair – the different technologies of tube and sheet hydroforming are described – differences in manufacturing practices required for aluminium compared with sheet steel are highlighted – the scope for superplastic forming of metals is considered and reference made to techniques used with plastics.

5.1 Steel formability

5.1.1 Sheet metal pressworking

The majority of parts comprising the bodywork of a current mass produced motor vehicle are shaped by pressworking, i.e. a sheet metal blank is made to conform to the required contour largely by a mixture of drawing and stretching within the initial main draw die. This first operation is illustrated in Fig. 5.1 which shows the cast iron main draw tool set mounted within the press frame. This is normally followed by

Fig. 5.1 2000 tonne main form press showing blank feeder mechanism[1]

four or five subsequent operations in tandem which in turn consolidate the features of the panel ('restrike') and then progressively trim the peripheral shape and pierce holes where required.

Loading the press can be done manually or by using a destacking and feeding mechanism whereby batches of sheet are delivered to the side of the press, the sheets magnetically separated and delivered into the jaws of the press by transfer using rubber suckers or a sliding roller mechanism. Lubricant can be pre-applied electrostatically or by spraying on entry to the press.

This type of tandem line was popular 20 years ago but in more recent years massive investment has been made in enclosed progression or tri-axis presses of the type shown in Fig. 5.2. Four successive operations are carried out within a massive enclosed frame, the parts being moved from one station to another by a walking beam transfer system. Typically, a Hitachi Zosen 5000 tonne cross bar feed press operates at speeds up to 15 strokes per minute and die changes can be effected in 5 minutes, and the target running efficiency is in excess of 70 per cent. The running of such an installation[1] calls for exceptional cleanliness and accuracy and effects such as 'pimpling' due to the impression of small particles of atmospheric debris or zinc from cut edges of coated steel blanks can cause serious problems due to the high number of panels produced before detection (often not until the painting stage) and on general productivity.

As this is the type of press that will feature more often in the future it is worth reiterating precautions that have to be taken to run such a line efficiently:

- More efficient and uniform packaging from suppliers
- Controlled processing through blanking operations with optimized die clearances
- Stillages in stainless steel
- Environmental controls to ensure the pressure inside the press shop is maintained above that of adjoining facilities
- Washing equipment incorporating amorphous filtering capable of removing particles down to 5 micrometres in diameter
- A second washing operation as blanks are automatically fed into the installation

Fig. 5.2 General view of modern tri-axis press installation

Although most of the emphasis of this chapter is concerned with material properties and the relationship with behaviour during component manufacture and how they influence performance, it is nevertheless worth recapping the effect of lubrication during pressing as this has such a profound part to play in performance. Systems used for tandem operations mainly relied on mill-oil plus manually applied high performance paste and recent campaigns sought to replace these with selective spray application. However, this type of system can be unsatisfactory from a housekeeping viewpoint, due to drippage onto the floor of the press bay, so electrostatic or pre-lubricant application of wax films was developed but even at coating weights of 2–3 g/m^2 these still tended to attract particles. It has been found that for tri-axis operation the more accurate tool location associated with this type of new installation, general uplift in washing procedures and slight enhancement of wash-oil lubricity plus chrome plating of critical tool surfaces has enabled most jobs to be run without interruption. Careful monitoring of mechanical properties also ensures that the increasingly high levels of drawability and stretchability, now a feature of today's steels, are fully exploited for each specific job. The more fundamental aspects of the interaction of surface with lubricity will be considered later, together with surface topography.

At the outset it should be stated that much of the extensive research carried out on the formability of sheet materials is laboratory based and therefore conclusions reached must be qualified by differences in geometric scale, punch velocity, etc. encountered under normal operating conditions in the press shop. Although facilities have improved considerably in recent years, performance will also be subject to operational changes due to constant reworking of tools, minor changes in tool/press alignment and locational factors arising from the change in press condition/location. Thus the sheet must have a wide tolerance between wrinkling, the unacceptable condition that develops due to the material drawing in too freely at the blank edges, and splitting/necking due to plastic instability. Above all, however, and this will be a recurrent theme in most other chapters, is the need for product consistency. This is emphasized in a classic paper by Butler and Wallace.[2] Even with moderate properties the press can be set to run continuously and maintain a high level of productivity, but if material properties fluctuate from very good to very bad then frequent changes become necessary with attendant frustrations which further add to downtime and variable quality.

Improvements in press monitoring techniques and 'fingerprinting' of specific jobs in presses using strain gauges mounted at the four corners of the press to reinstate tools to previous settings, have assisted in reduction of setting-up time and minimizing adjustments. Statistical process control of material properties has also enabled a database of control parameters to be established by recording range and average values, with feedback to material suppliers to tighten appropriate parameters and achieve increasing uniformity. Where a number of suppliers are involved it is often worth comparing performance using weighted values for selected parameters, so that comparative ratings can be derived and relative positions in the 'league table' presented (anonymously) at regular review meetings. Thus running conditions and material variability have generally improved but consistency remains a basic requirement.

A controversial issue has always been the choice facing the user of purchasing material to 'make the part', whereby he can claim replacement material if the job proves problematic, or alternatively procuring material to a detailed set of properties, i.e. 'specification buying'. As running conditions and accuracy of tool location, etc. improve, rejection to tighter limits is now required by the pressworker and supply to a detailed specification including strength, surface and dimensional and forming parameters is becoming the norm. With modern steelmaking practices described in Chapter 3 it should be no problem for the supplier to respond and tailor material to the values prescribed for individual jobs but a quick turnaround of replacement material, if required, is essential to prevent undue disruption of pressing schedules. It is therefore paramount that calibration of supplier and user test facilities is carried out on a regular basis and any margins for error agreed, so that any disputes concerning properties may be speedily resolved. Reject coils can occupy valuable space in any operation run on a 'just in time' basis and for maximum logistical efficiency must be removed quickly. To promote speedy material movement the setting up of supply satellite warehouses near the plant is to be encouraged. As well as tensile parameters, which will be defined later, physical attributes must also be monitored. The case for increasingly accurate dimensional control has been made in the past[3] and international standards showing thickness tolerance ranges of plus or minus 6–8 per cent against a capability to roll within 1 per cent (typical of most modern mills) for cold rolled materials, must be open to question.

Consistency of running is as equally important as consistency of material supply. Thus it is important to achieve the same performance in running as when the initial tooling was developed to maturity. A fairly rigid procedure must be adhered to by which the material grade is selected at the concept feasibility stage of a model programme and followed through at all stages of tool development. Ideally when material is ordered for the die 'try-out' stage this should carry explicit instructions that properties should reflect those at the lower end of the thickness and property band so that the tools can be worked to accommodate this material and hopefully then establish a realistic tolerance window for operation in the future. The danger is that tooling can be outsourced for manufacture and during 'buy-off' an acceptable performance is established using an exceptional grade of material. On initial production runs it is then found that the job is impossible to run or that high scrap rates are incurred and therefore it is imperative that a detailed record of material properties used for each of the development stages is kept for inspection, otherwise unrealistic properties will result in elongated downtime and prolonged production runs.

With this background it is now timely to introduce the individual parameters relevant to sheet materials during press forming and how they are measured. The concept of 'forming limits' will also be used to explain the relevance of each of the parameters and how these can be used to assess criticality and improve performance. As stated it is essential that close co-operation exists between supplier and user on a technical basis so that an understanding of the requirements for each job are quickly agreed. A description of test methods and procedures commonly adopted now follows.

5.1.2 Sheet properties and test procedures

Although most shapes made by press forming are produced under biaxial conditions involving complex strain paths, uniaxial test methods are invariably used for the routine monitoring of properties, the type of test piece and parameters commonly referred to being described below. The test piece is of the general type shown and the load applied to failure using a universal type tensometer which has been calibrated against national standards. Most press shop test houses also carry out periodic checks against facilities used by suppliers to ensure meaningful comparisons can be made on similar samples and that material falling outside the users' specification is not shipped from the steel mill. Occasional use is made of biaxial test methods but these require methods such as bulge testing for which test procedures are difficult to standardize/cross-correlate and involve slow measurement of stress and strain.

5.1.2.1 Parameters derived in the uniaxial tensile test

The tensile test, carried out to prescribed test methods such as BS EN 10 002-1:1990, provides information on the following parameters, which are illustrated using a stress/strain diagram such as that shown in Fig. 5.3.

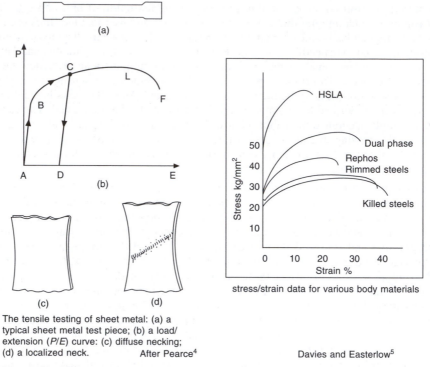

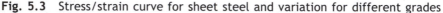

The tensile testing of sheet metal: (a) a typical sheet metal test piece; (b) a load/extension (*P/E*) curve: (c) diffuse necking; (d) a localized neck.   After Pearce[4]   Davies and Easterlow[5]

Fig. 5.3   Stress/strain curve for sheet steel and variation for different grades

The test piece normally used for monitoring sheet properties has an 80 mm gauge length and typically a microprocessor controlled 200 kN capacity test machine is used to automatically deliver test results and print an engineering stress/strain curve of the type shown above.

The form of the curve is normally smooth, exhibiting an elastic portion where stress is proportional to strain followed by a plastic zone where work hardening proceeds failure at the ultimate tensile stress. The accompanying elongation can be in excess of 40 per cent for forming grades reducing to less than 20 per cent for some high strength grades.

The yield or proof stress indicates the initial strength of the material and the associated yield point shown at the limit of proportionality can be smooth or discontinuous in the case of aged or bake-hardened material (see Chapter 2). Associated stress levels usually vary from 140 N/mm^2, the accepted design strength for forming grades, to 300 N/mm^2 for high strength body panels. This may rise to 800 or even 1200 N/mm^2 for selected parts which are safety related, e.g. door intrusion rails.

As a general rule, elongation normally falls with increasing strength level on a reasonably smooth curve but exceptions are now emerging where the strength is disproportionally high. This occurs with dual-phase and TRIP steels (discussed in Chapters 2 and 8). Elongation is a measure of ductility and where no necking can be tolerated uniform elongation should be taken as the true comparator. To further assess drawability, '*r*' value, which is a measure of 'thinning resistance in the thickness direction of the sheet', is taken. A high '*r*' value is indicative of a sheet with superior drawability while a large difference between '*r*' measured in the 0°, 45° and 90° directions would be prone to 'ears' forming at the flange periphery, requiring trimming and wastage of material.

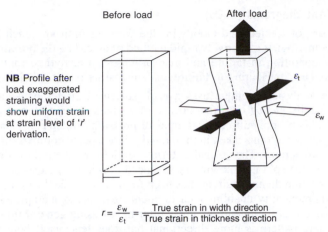

NB Profile after load exaggerated straining would show uniform strain at strain level of '*r*' derivation.

$$r = \frac{\varepsilon_w}{\varepsilon_t} = \frac{\text{True strain in width direction}}{\text{True strain in thickness direction}}$$

Fig. 5.4 Derivation of '*r*' value for sheet metals

For practical purposes measurements are taken of width and length and the '*r*' value computed from the following equation, assuming constant volume:

$$r = -\ln(W/W_0)/[\ln(L/L_0) + \ln(W/W_0)],$$

where W_0 and L_0 = original width and length, respectively.

Work-hardening exponent '**n**'

This parameter is derived from the plastic strain portion of the stress/strain curve and provides a measure of stretch-formability. More specifically it is calculated from the slope of the true stress/true strain plot as shown in Fig. 5.5.

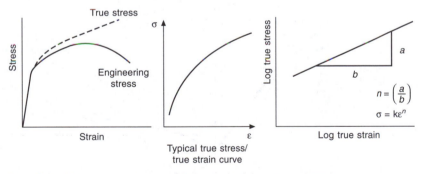

Fig. 5.5 Work-hardening coefficient (*n*) from stress/strain curve

True stress/true strain data is of more relevance to engineers and structural analysts, and is increasingly requested for input to CAE and FEM forming and impact simulation programmes. However, the more commonly encountered 'engineering stress/strain' curves are more easily constructed and more widely compared in supplier–user technical discussions. True strains are additive[9] and so the following applies:

$$\varepsilon_1 + \varepsilon_2 + \varepsilon_3 = 0$$

where ε_1 = major strain

ε_2 = minor strain

ε_3 = thickness strain

5.1.2.2 Forming limit diagrams (FLDs)

Since the original work of Keeler[6] and Goodwin[7] the forming limit approach has enabled press shop technologists a relatively rapid method of assessing the proximity of a measured strain condition to failure and also provides an analytical tool for prioritizing the influences contributing to failure and suggesting possible solutions.

Essentially the strain analysis technique uses a grid pattern of circles or squares which is electrochemically etched onto the surface of a blank using a stencil and suitable electrolyte.[8] This is applied to critical areas of pressings which can then be formed and measurements of strains taken in major and minor axes to compare either strain gradients or failure strains. The normal criterion for defining the failure strain is 'the nearest ellipse to the split not showing pronounced necking'. A close-packed hexagonal pattern of 2.5 mm diameter circles has been found to provide the optimum blend of accuracy and practicality, enabling measurements to be made in all principal directions while giving a useful guide to the press fitter in assessing general flow of the material. The square pattern is more directional but does lend itself better to automatic strain measurement using video camera and associated techniques. Whatever the pattern precautions should be taken to ensure the redeposited compound constituting the grid mark does not interfere with the lubricity of the forming operation and that the original mill oil/lubricant condition is restored after etching.

The forming limit curve (FLC) (Fig. 5.6) shows the position of failure for a range of different strain combinations tending to reflect drawing conditions on the left-hand side, characterized by a contraction in the minor strain axis and elongation in the major direction, and expansion in both directions for stretching operations as on the right-hand side.

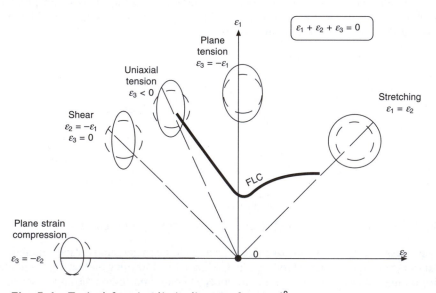

Fig. 5.6   Typical forming limit diagram for steel[9]

Using the highest major strain and associated minor strain enables the strain ratio to be identified quickly and the severity at that location to be established by proximity to the failure curve. Assessment of the safety margin may allow downgrading to be considered. Conversely, critical strains may be eased by working the tools to release material from adjacent features, using the flow pattern to trace the origin.

In this context the use of FLDs is not restricted to production troubleshooting and is regularly used in die try-out to agree a trouble-free condition has been attained before 'buy-off' or acceptance by the customer. Contributory influences which alter the height of the curve are thickness and 'n' value and this should be considered when comparing strain values.

5.1.2.3 Modelling of the sheet forming process

The numerical approach referred to above lends itself to automation of strain measurement as described above, one approach featuring two TV cameras taking accurate readings over a critical area of a gridmarked pressing and displaying measured strain ratios on an FLD for that material. An alternative (and cheaper) approach is to use one camera and take individual measurements of ellipses before reference to an FLD compiled for a specific material. This approach, although useful in troubleshooting, is retrospective and ideally a mathematical solution should be found to optimize die design so that the die try-out process is abbreviated by elimination of guesswork – necessary with even the most informative knowledge-based design system.

The complexity of the required system is, however, immense as the focus of forming operation is a thin-shell (rather than a solid) form, the material behaviour is non-linear, and any model must accommodate any differences in anisotropy, work hardening, friction (workpiece and tooling surfaces, lubrication), springback and other processing parameters, all of which can change during deformation. The required information includes all aspects of try-out including necking, splitting, wrinkling and buckling. However, significant progress has been made in the last 60 years in the development of non-linear plasticity theory[10,11] and finite element methods[12,13,14] to now start to apply such techniques with some confidence and, most importantly, to complex pressings, rather than axisymmetric shapes. Sheet forming simulation groups are now a permanent feature of any major automotive organization and can provide fairly rapid 'one-step' solutions which assume linear strain paths and are based on the initial and final configurations. These make drastic simplifications and for a more detailed analysis of die feasibility the commercially available LS DYNA-3D, PAM-STAMP, or OPTRIS FEM simulation codes are used. As well as 'in-house' automotive support such feasibility studies are often used by suppliers such as Corus as part of their customer service operations.

5.1.3 Effect of surface topography

The different types of surface finish applied to steel and aluminium have been described in Chapter 3 but a brief reference to their effects is now necessary. Early experimentation in the 1960s and observations[15] suggested that two major parameters were significant and could be routinely measured in the press shop, namely, peak-count, P_0, and centre-line-average (CLA) surface roughness, R_a. Using stylus type contact measurement techniques values could be consistently reproduced in the test house alongside the steel mill and subsequently compared with readings at the pressing plant. Further regular press shop investigations through the 1970s suggested that for internal parts a roughness range of 1.0–1.8 microns together with a peak-count of 80 peaks per cm provided the best compromise between oil retention during forming for optimum lubricity and an adequate finish. For external parts where a critical paint finish was expected it was shown that a lower CLA of 1.0–1.5 microns was required together with a higher peak density of 100 peaks per cm. The rougher finish helped to disguise score marks due to localized cold welding at tooling/blank high spots while the smoother finish helped improve image clarity of the paint film. These two parameters are still the most quoted control measures and a European standard has now been agreed for their derivation, EN 10049.

Later comparison of steels on a worldwide scale[16] in the mid-1980s using a 3-D stylus technique appeared to show that steels with a plateau type surface profile, i.e. predominance of peaks with a few valleys to retain lubricant gave a more consistent press performance. It was also observed that steels with a longer wavelength, i.e. peak to valley profile measured over millimetres rather than microns, tended to show more of an orange-peel paint finish. Thus the distinction was made between macro- and micro-surface profiles to control overall flatness and image clarity. This highlighted the importance of the last roll in the tandem mill and the temper mill work rolls in achieving these finishes – and the limitations of sand blasting in developing and maintaining these profiles.

The last decade has therefore seen the emergence of the various profiles referred to in Chapter 3 typified by the choice of Sibertex finishes shown on p. 82. Thus in terms of control the trend to more deterministic finishes must be an improvement in the press shop and advantages of specific treatments should become more evident during painting.

The various profiles are summarized in Fig. 5.7, extracted from a recent BMW paper[17] describing a detailed programme to determine the effects of topography, which is considered to have an effect on geometric accuracy, suitability for joining (bonding, welding and bonding), adherence of metallic cladding and organic coatings as well as the forming and painting influences mentioned above.

The parameters studied included:

• Total height of profile

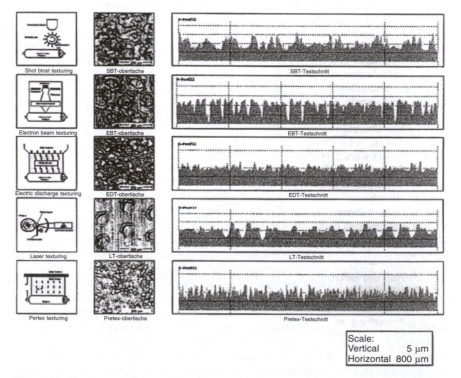

Fig. 5.7 Various surface profiles developed for automotive steel sheet. (After Kopietz *et al.*[17])

- Oil volume
- Contact surface
- Profile peak height
- Skewness (bias above or below centre-line)

It was shown that sheets with a higher degree of roughening showed the following benefits on forming:

- Reduced anisotropy of properties
- Better geometrical accuracy due to a higher static friction component
- Increased forming capability
- Better adhesion of coatings and cladding

As referred to elsewhere, these finishes should be universally available if maximum consistency and performance are to be obtained.

As apparent from above, the subject of sheet metal forming interactions ('tribosystems') is extremely complex but an excellent review of the tribological effects of various parameters such as substrate type, coating type and topography on frictional characteristics was presented by Schey[18] and this is recommended for further reading. Regarding material types discussed here, a study of high strength steels was reported in this review which showed friction to fall with increasing hardness. Referring to zinc coatings, IZ (iron–zinc alloy) was found similar to bare mild steel, the same conclusion applying to electrozinc, but the COF (coefficient of friction) for hot-dip coatings was generally lower and decreased with the number of samples drawn in succession in the strip-draw test (a popular method of comparing frictional characteristics).

5.1.4 Effect of zinc coatings

In the author's experience IZ shows a similar performance to mild steel sheet, the lubricity being enhanced slightly by absorption and retention of the lubricant in the fissures typical of the coating. IZ is notorious for appearing to be deficient in mill applied lubricant but coating weight tests suggest it is present, an effect attributed to the absorbent nature of the network of cracks. The natural tendency of the fissured structure is to powder and even flake – although recent modifications to the substrate (Nb/Ti now being preferred to Ti) and optimization of thickness ($45\ g/m^2$ recommended) have improved the tendency for 'pimpling'. This effect, when zinc-rich particles deposit on the punch and are impressed through the sheet to give a shallow mound, only shows on painting or stoning which means a high number of panels requiring rework are produced before the defect is recognized. This is now less of a problem with EZ coatings as process disciplines, which reduce particle generation, e.g. use of side-trimmed strip, regular die cleaning and blank washing, have been progressively introduced. FLDs have also been constructed in an attempt to predict coating behaviour under various strain regimes. Reverting to coating lubricity, electrozinc coatings have been slightly beneficial but the press performance of drawn parts such as spare-wheel wells and door inners has definitely been improved by the use of hot-dip coatings, although tools should be inspected regularly for signs of pick-up.

5.1.5 Tooling materials

As sheet materials have developed in recent years, so too have the materials used for the production forming tools. This is due to two main reasons:

- A general recognition that for consistency of operation and quality, stable materials and controllable process treatments were essential. Past emphasis was on higher volumes and avoidance of interruptions to production, ease of repair, preferably in the press, was essential, and lower cost water hardening cutting steels were ideal. Of simple composition these could be 'feathered in' at the press and production resumed quickly. However, the repairs were often unsatisfactory with welding imperfections and distorted profiles. This view has changed more recently and even in high volume situations it is preferable to stop and repair properly to achieve a consistent condition – or better still to use a higher quality material (oil- and air-hardening steels) to start with.

- The use of higher strength blank materials which require a higher press load and consequently accelerate tool wear. As well as prompting a search for more effective materials this has led to the development of surface and heat treatment.

The preferred tooling materials in current use for production of a typical body content as illustrated in Table 2.1 are summarized below under main form, cutting tools and surface heat treatments.

5.1.5.1 Main form tools

This category mainly refers to the larger 3-D tools used for the production of the main shape of the panel and may extend to large panels such as bodysides and floors. These form the first set of tools in a series of five or six operations and weigh up to 50 tonnes. Presses may be single or double action, but in either case separate punch and blankholder control is essential to allow flow of the sheet material which can exhibit strain levels of up to 30 per cent in stretching operations and more than 100 per cent in major strains during drawing. Hence for the large areas of 3-D surfaces with major draw depths and contours associated with wheel arches, door inners (often produced as double pressings), and bodysides a large area of the tool surface may be subject to sliding wear with peripheral zones in particular operating under severe pressure. The main requirement therefore is that this relatively large mass should be easily shaped, show good frictional properties and high resistance to both compressive and impact loads. Machinable cast irons are a good choice and common selections are shown in Table 5.1 together with the main cutting steels.

To maximize lubricity these are normally flake irons (some SG irons being specified in specialized applications) within a pearlitic matrix to maximize wear resistance. For tooling where severe wear resistance is required, special alloy formulations are used containing chromium and molybdenum which render them suitable for flame hardening or laser hardening (less distortion and local damage) in critical areas. Alternatively, these areas can be specially surface treated by plasma nitriding or hard chrome plating.

Cast irons are also used for aluminium although for main form operations cast steel is gaining in popularity as it is less prone to pimpling effects and small surface blemishes which have been attributed to cast irons in critical situations. Good wear resistance is also obtained from aluminium bronze tooling.

5.1.5.2 Cutting materials

Most operations following the main form and restrike operations (used to improve definition of features) used to either trim the shape or pierce holes. Therefore wear and impact resistance are essential and medium carbon and alloy steels are usual choices. Historically, water-hardening plain carbon steels were widely used due to low cost and repairability in the press, using fairly crude heat treatments.

Table 5.1 Typical tooling materials used in main form and cutting operations

Material type forming	Designation: DIN/AISI	Use	Hardness HB	Additional comments
Grey flake cast	GG 25	Main form tooling	160–210	Low cost; localized wear 'soft spots' may form at ferritic zones.
Alloy flake cast	GG25CrMo	Form and draw	210–250	Alloying promotes more uniform pearlitic structure. Flame hardenable for added wear resistance.
SG iron	GGG 50	Cam operations	150–180	Graphite developed in spheroidal form for improved toughness.
Alloyed SG cast	GGG 70	Form and draw	240–270	Spheroidal graphite nodules impart additional wear resistance.
Steel castings	GS 60, 70	Form and draw	175–252	More uniform structure for improved surface and wear resistance.
Cutting steels. Oil-hardening tool	AISI type 01	Cutting and trim	229	Low alloy steel with high surface hardness and toughness, after hardening and tempering.
Medium alloy tool	AISI type A2	Cutting and trim	240	Good combination of shock and wear resistance.
Air-hardening chromium steel	AISI type D2	Cutting and trim	250	Excellent wear resistance, requires careful heat treatment.

However, it has been found that performance could be erratic and that improved properties and stability of form could be obtained from the use of oil-hardening steels of the AISI 01 type and in cases of exceptional wear high chromium (12 per cent) air-hardening D2 steels. These require accurate control of conditions during heat treatment, and investment in state-of-the-art salt bath, fluidized bed and furnace facilities is essential within the modern press shop if high productivity and quality levels are to be achieved. Facilities should include salt bath and fluidized bed equipment together with necessary control instrumentation and material test equipment.

As with castings, surface treatments improve wear resistance and for smaller cutting tools ion implantation chemical vapour deposition (CVD) and similar treatments, whereby surfaces are hardened to a depth of less than 0.01 micrometre, have been claimed to increase tool life by many times. An excellent review of this subject has been presented by Bell.[19] One aspect that must be considered, however, is repair, as incidental damage by chipping, etc. cannot be as readily rectified as normal steels by welding and heat treatment.

One method of producing a cutting edge which combines the economy of cast iron with the performance of steel is hard edge welding and may be the most suitable for larger cutting application. This comprises preparation of a chamfered edge on the cast iron tool and the progressive laying down of 'buttering' intermediate layers of nickel to provide a cushion for keying to the substrate followed by the final layer of cutting material. This is finally machined to provide a wear and shock resistant combination at low cost.

5.1.6 Hydroforming

5.1.6.1 Tube hydroforming

The design and engineering aspects of hydroforming have been described in detail in Chapter 3. It is clear that the overwhelming majority of autobody structures are manufactured using pressed steel parts subsequently spot welded together to form a monocoque. Primary load paths are via the box section geometry. An alternative method to achieve similar box section design is through the use of hydroformed tubes, with a number of potential advantages.

Tube hydroforming is a process whereby a section of fabricated steel tube is inserted into a specially manufactured press tool. The tool is closed and the ends of the tube are capped. A fluid is then introduced into the tube under high pressure and as the pressure is increased the liquid inside the tube forces the tube walls to expand outward to conform to the shape of the tool. End feeding of the tube material into the tool may take place to reduce thinning. Finally, the pressure is removed and the part is removed from the tool. Subsequent fabrication processes of cutting and welding can be used to finish the component. The appearance of the tube at each stage of the process is illustrated in Fig. 5.8.

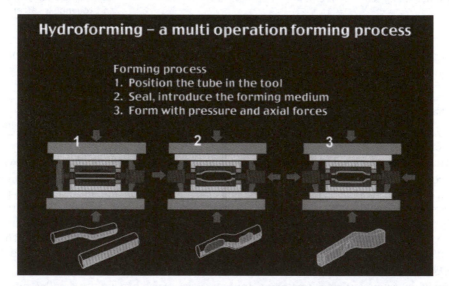

Fig. 5.8 Tube development during the hydroforming process. (After Tertel[20])

In the present context, the hydroforming process raises a number of interesting materials engineering issues.[21] The actual tubemaking process itself can induce an increase in the strength of the steel strip to varying degrees around the circumference of the tube. A resulting loss in ductility or n value can occur, which for a stretch forming operation such as hydroforming can be critical. Unlike the traditional stamping operation which induces highly variable levels of strain (and hence strength) in a component, in the hydroforming process the strength development and consistency offered may allow the use of a lower grade and strength material. However, thickness effects need to be considered as well. Traditional practice for engineering with stamped components is to assume thickness reductions due to thinning will balance the increase in strength due to strain hardening. In the case of hydroforming, material will thin in the absence of end feeding during the hydroforming process. Significant strengthening due to deformation during hydroforming is a feature of hydroforming but can reliance be placed on this for design purposes? What is the relevance of FLDs largely derived during sheet metal stamping operations in assessing the criticality of hydroformed parts? Also requiring consideration are the changes in the material properties that can result from interstage heat treatments in the hydroform process (often used to maintain a certain level of ductility). If the strain levels in the material are at a certain level, resulting in critical grain growth on annealing, this may even result in lower strength levels than those in the parent material. There are many aspects to materials technology within the hydroforming process that require fuller resolution and key issues are highlighted below. Nonetheless, the use of hydroforms in the BIW appears to be developing with recent applications by General Motors and BMW employing hydroforms in production models and Volvo and Land Rover examining their use in concept vehicle studies.[22]

It is interesting to note that tube manufacture in diameter-to-thickness (d/t) ratios required by designers can be problematic. Traditional precision tube supply has been based on upper d/t ratios of approximately 60, while ratios approaching 80 or more may be desirable for automotive hydroforming. Continuous tube manufacturers continue to work on developments in this area in the expectation of increased use of hydroforms for autobody applications. The above ratio restrictions do not apply to tube press formed from a sheet blank then laser welded.

Characteristics of hydroformed parts

Thickness effects. In the absence of significant end feeding during hydroforming, it should be noted that the material will thin. This reduction in thickness must also be taken into account in predicting the structural performance (static and dynamic) of hydroformed components. This is in contrast to the traditional practice for stamped components where changes in thickness and strength have largely been assumed to balance and as a result have been ignored. It is likely that hydroformed components will undergo more widespread thinning than the stamped components they replace. Conversely, where significant end feeding is possible during hydroforming, the final thickness may be at least equal to the starting value and this, together with an associated increase in yield level, may result in a component that is considerably stronger and more rigid than anticipated. If the component is a longitudinal crash member then this may transfer collapse, undesirably, further rearwards into the vehicle. In such a case appropriate design allowances may have to be made. There is, therefore, a need for rigorous formability simulation methodologies to predict the through-process strength and thickness changes: such analyses should ideally be amenable to being fed through to static and dynamic structural performance software. This is a current area of active research worldwide.

Forming limit diagrams. The complex strain history which hydroformed components may undergo, due to a necessity to introduce a prebending and preforming operation prior to the actual hydroforming stage, is illustrated in Fig. 5.9.

During a detailed practical investigation[21] it was found, first, that the forming limit curve for the tube was somewhat lower than the parent coil (especially in the position 180° opposite the weld) and, second, the major strain at position 90° was observed to reduce from the first to the second stage, and to reduce yet again in the final

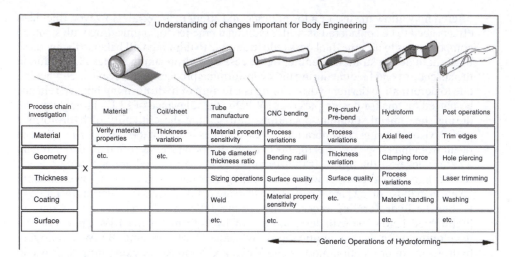

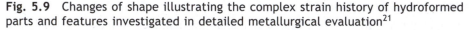

Process chain investigation	Material	Coil/sheet	Tube manufacture	CNC bending	Pre-crush/ Pre-bend	Hydroform	Post operations
Material	Verify material properties	Thickness variation	Material property sensitivity	Process variations	Process variations	Axial feed	Trim edges
Geometry	etc.	etc.	Tube diameter/ thickness ratio	Bending radii	Thickness variation	Clamping force	Hole piercing
Thickness X			Sizing operations	Surface quality	Surface quality	Process variations	Laser trimming
Coating			Weld	Material property sensitivity	etc.	Material handling	Washing
Surface			etc.	etc.		etc.	etc.

Generic Operations of Hydroforming

Fig. 5.9 Changes of shape illustrating the complex strain history of hydroformed parts and features investigated in detailed metallurgical evaluation[21]

hydroformed component. Each of these changes was accompanied by a corresponding change in strain path. The adoption of a forming limit curve derived in the laboratory using a test method involving an essentially linear strain path was therefore considered highly questionable. There is sufficient data in the published literature to demonstrate that a forming limit curve determined using a complex strain path can be quite different to that for a linear strain path. The trend is generally for the failure strain to increase if an initial uniaxial tension mode of strain is followed by a biaxial tension mode (i.e. both the major and minor surface strains finally become positive). Conversely, initial biaxial straining followed by uniaxial deformation to failure can decrease the failure strain relative to that determined using a linear strain path throughout.

More recently, interest has been growing in the adoption of the concept of a forming limit stress curve (FLSC), which is claimed to be essentially independent of strain history. Here, the strain FLC is changed to a corresponding FLSC by converting from strain to stress using the work-hardening (stress/strain) relationship for the required material. With reference to the FLSC, this indicates the failure stress ranging from the uniaxial tension state (zero transverse stress) to plane strain (transverse stress approximately equal to half the longitudinal stress, depending on material anisotropy) and, ultimately, equibiaxial straining (longitudinal stress = transverse stress). Regardless of the strain history of the forming process, the material is deemed to have attained its failure limit when the state of work hardening reaches the FLSC level. The state of strain in a component may first be determined using circle grid analysis techniques, or FEA simulation, followed by conversion to equivalent strain in order to calculate the equivalent stress.

Cold work strengthening. The considerable increase in yield strength resulting from the hydroforming process prompts consideration of using such a phenomenon to achieve, say, weight saving and/or enhanced structural performance. By way of illustration, strengthening from cold work is currently used in other market sectors such as cold formed sections for the construction industry. Further, the IISI/ULSAC project proposes the use of sheet hydroforming to promote increased (and uniform) stretching in order to maximize the dent resistance of closure panels through increased cold work.

In each of the above instances, the in-service structural requirement is essentially one of resistance to quasistatic, as opposed to dynamic or constantly fluctuating, stresses. In the latter case, there is sufficient evidence elsewhere to show that the imposition of high levels of cyclic stress (or strain) can give rise to the phenomenon of cyclic softening, thereby largely negating the strengthening effect due to cold work. Data generated by Corus during a separate study on the influence of prior cold work on low cycle fatigue performance is summarized in Table 5.2.

Table 5.2 Low cycle fatigue of 10% biaxially prestrained sheet steels[21]

Grade	Monotonic yield	Stress (N/mm^2)	Test condition	Cyclic strain(%)	Cyclic stress at half-life (N/mm^2)	Reversals to failure
	As received	After 10% biaxial strain		+/-	+/-	(2Nf)
EDDQ/AK	159	315	As received	(2Nf)	126	424 158
			10% biaxial			
			As received	0.2	163	82 036
			10% biaxial	0.2	221	11 440
HSLA	321	493	As received	0.1	207	845 868
			10% biaxial	0.1	210	940 178
			As received	0.25	298	27 144
			10% biaxial	0.25	324	6 402
Dual phase	369	726	As received	0.1	187	>2 700 000
			10% biaxial	0.1	201	>3 000 000
			As received	0.2	326	59 134
			10% biaxial	0.2	380	29 154
			As received	0.25	342	28 506
			10% biaxial	0.25	446	10 336

Three different types of steel were examined, ranging from low carbon mild steel to dual phase. The trends were similar for each: the prior 10 per cent biaxial prestrain gave rise to an appreciable increase in monotonic yield stress. However, cyclically loading at constant strain amplitude levels within the yield region resulted in a cyclic stress level (at half-life) that was well below the monotonic yield after prestraining. Not all the benefit in cold work strengthening was lost because the cyclic stress levels were all higher than the corresponding stress levels for the unstrained (as-received) condition. Further, the fatigue life of the mild steel after prestraining was, at all cyclic strain levels examined, lower than for the unstrained condition. For the other two steels, the fatigue lives of prestrained samples were lower when fatigue tests were carried out at ± 0.2 per cent strain and above, but the reverse was the case at ± 0.1 per cent possibly because this level of strain was insufficient to induce significant levels of cyclic plastic strain. Additionally, if the tensile strength is only marginally increased through work hardening then dynamic properties such as high cycle fatigue life and impact resistance may be similarly only slightly enhanced. As a result of this work it was felt that there was a need for further understanding of the effects of strain hardening on fatigue of components.

In conclusion a degree of caution is recommended when assuming experience with stamped steel parts, which can be extrapolated to equivalent forms produced by tube hydroforming, especially with regard to interpretation of FLDs, thickness development and design calculations based on the expectation on the full extent of increased yield stress values induced through deformation. It is emphasized that any hydroforming process requiring interstage annealing rules out the use of zinc coated finishes unless recoating by spraying or an alternative process is a possibility.

5.1.6.2 Sheet hydroforming

Large panels such as roofs, bonnets and doors can, depending on design criteria such as curvature, exhibit low buckling resistance at the centre of the panel. This is commonly caused by the low levels of strain in the sheet, which results in low levels of strain hardening and low strength. As described in Chapter 3, this can have a detrimental effect on panel dent resistance. As a result higher strength materials or reinforcing components may be added, also adding weight to the final body. Sheet hydroforming is a relatively new technology offering the potential to overcome this difficulty. Advantages include:

- Improved component strength and stability (dent resistance and improved buckling through slight change in shape)
- Excellent surface quality
- Potential to process different materials and thicknesses with a single set of tools

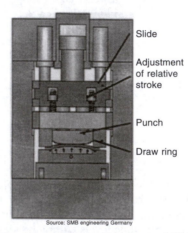

Source: SMB engineering Germany

Fig. 5.10 Sheet hydroforming process. (Courtesy of ULSAB Consortium)

The process begins by preclamping a large blank with a blankholder and a watertight seal. The liquid emulsion is introduced and pressurized behind the blank. This causes a controlled bulging of the blank with subsequent work hardening. Following this prestraining operation the pressing operation commences defining the shape of the component. The ULSAB programme employed the sheet hydroforming process for the production of the vehicle roof.

5.2 Aluminium formability

The automotive industry has been using aluminium sheets for low volume applications for many years but by and large there is still a lack of understanding both in the design office and on the shop floor with respect to the behaviour of the material. When difficulties arise there is a tendency to resort to steel because it is better understood and there is a wealth of experience that can be drawn upon. In spite of efforts on the part of aluminium suppliers and many research teams around the world, forming aluminium panels still remains a challenge and the potential problems that may occur during production are varied. Due to 'bolt-on' assemblies being the easiest method of achieving significant weight savings at moderate cost and with minimal assembly problems, the initial emphasis for many manufacturers has been on outer panels. Hence, particular attention will be given to the special requirements of skin panels although much of the technology relates to all applications. Most of the general backgound presented above on the relevant parameters and test techniques applicable to the formability of steel also applies to aluminium and the emphasis here will be on specific aspects which characterize the properties of this material in sheet form.

Uniaxial engineering stress/strain diagrams for comparing steel with aluminium are shown in Fig. 5.11.

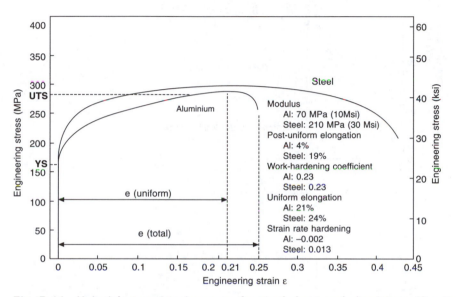

Fig. 5.11 Uniaxial stress/strain curves for steel sheet and aluminium. (Courtesy of the Aluminium Association[23])

Aluminium being less formable than steel presents a challenge in obtaining a tight skin panel that meets dent resistance and shape stability requirements. Achieving target strains of 1–2 per cent in the centre of the panel during the drawing or first forming operation often leads to splitting in peripheral areas. When such target strains are not attained, the panel will not exhibit the dent resistance required and will likely present areas of 'loose metal' (i.e. areas where the panel can easily depart from its intended shape). In addition, there is a definite tendency to galling during forming in the conventional tools for deep drawing. This signifies that maintenance of the draw dies is an important issue if scoring of the panels is to be avoided.

Comparing the FLD for typical aluminium alloys used for sheet production in the automotive industry with that for steel (Fig. 5.12) shows that the critical strain levels of most of these alloys are inherently lower than that of steel. It has been argued that this does not necessarily imply poor formability as if the strain can be spread more uniformly by sympathetic die design and lubrication systems then similar shapes can be produced. It is generally accepted, however, that allowances have to be made for the fact that deep and complex automotive panels, e.g. clam-shell bonnets and multi-featured monosides, are often beyond the capability of aluminium alloys. This does not mean that it is excluded from the design of these panels but that alternative techniques may have to be employed to construct more complex fabrications.

The FLD for typical sheet alloys is shown in Fig. 5.12 together with a comparative curve for steel.

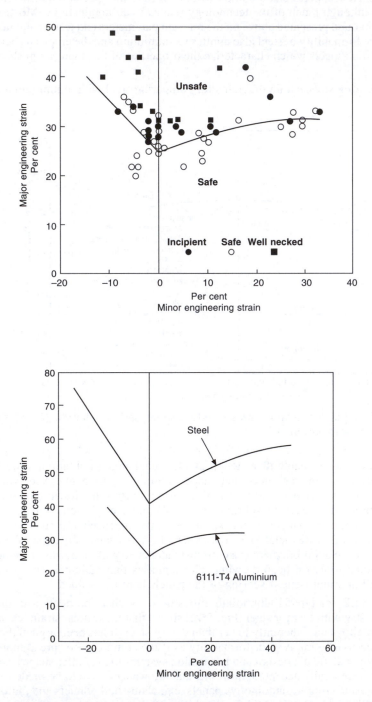

Fig. 5.12 Typical forming limit diagrams for aluminium compared with that of forming steel. (Courtesy of the Aluminium Association)

5.2.1 Simultaneous engineering approach to design with aluminium

It is well known that the quality of a stamping (or subassembly) is dependent on the design of each stage of the manufacturing process. Moreover the design of each manufacturing operation in the production cycle has an influence on the quality and success of subsequent operations. In other words, problems in one stage of forming may be the result of a poor design several stages upstream. Therefore it can be easily understood that there is interdependence between the various forming processes and stylists, product engineers, process engineers and toolmakers must work in closer collaboration in order to produce world-class quality panels and vehicles. In this section, based on a paper by Green *et al.*[24] following a comprehensive collaborative study on this subject, this simultaneous engineering approach to designing outer body panels will be described to resolve the particular issue of hemming.

5.2.1.1 Material selection

Hemming is a subassembly process that is commonly used in the automotive industry, when very compact and stiff joints are required and where the appearance of the edges has to meet the highest standard. In this assembly process, the flange of the outer panel is bent 180° around the edge of the inner panel. It is the experience of many manufacturers, however, that hemmed subassemblies using aluminium outer panels can present real difficulties if the expected bending radius is less than the thickness of the sheet. Due to the limited ductility of aluminium alloys the hemming process may lead to cracking of the outer surface and splitting along the bend line. In this case the complete subassembly is rejected and not only can the costs associated with this scrap rate be considerable but the production rate may also decrease significantly.

The first step in resolving the issue of surface cracking during the hemming operation is to involve the material suppliers. It is of paramount importance to understand the capabilities of the material and to design according to its specific mechanical properties. An aluminium stamping cannot be styled and designed in the same way a steel panel can. During the initial part of this collaborative programme, a detailed audit of each of the aluminium suppliers' production and testing facilities led to a clearer understanding of the manufacturing process of 5xxx and 6xxx series alloys. This also highlighted the fact that there is a definite relationship between the thermomechanical production process and the resulting microstructure, which in turn influences the bending performance of the material.

A substantial part of the programme was devoted to the mechanical testing of various 5xxx and 6xxx series alloys, which are currently used in the automotive industry. This testing was carried out with a view to determining not only the formability and anisotropy of the alloy but also its bending performance and bake-hardening response as well as surface topography and ambient stability. It is essential that styling, product and assembly design of aluminium panels is based on relevant experimental data and therefore it is strongly recommended that automotive manufacturers extensively involve their aluminium supplier at every stage of design.

One of the first decisions which have to be made before designing an aluminium outer body panel relates to the choice of grade and 5xxx and 6xxx series alloys are the most common automotive alloys, each offering specific advantages. Recapping on the earlier, more detailed reference, 5xxx series alloys are typically used in applications that require a high level of formability. They generally exhibit yield point elongation phenomena (coarse Lüders bands of type A, visible through paint) at low elongations (ca. 1 per cent), which make them unacceptable for 'Class A' (external finish quality) surfaces. This effect can be avoided by a properly designed thermomechanical production process (SSF = 'stretcher-strain free'). However, the fine Lüders lines of type B will arguably still be present at high levels of deformation (>5–10 per cent) after forming at room temperature. 5xxx series alloys also tend to soften (the yield stress will decrease) when subject to the paint curing cycle.

On the other hand, 6xxx series alloys (free of type A/B type bands) can harden significantly as a result of the paint-baking cycle. The yield stress will increase with prestrain, with the temperature of the ovens and the length of time the product spends in the ovens. Although 6xxx series alloys have been preferred for the production of outer body panels because of their superior dent resistance it is critical that the bake-hardening response of the material be evaluated under the actual time/temperature cycles in the baking ovens. The tendency nowadays is to reduce both the temperature and cycle time of the electrocoat oven and the actual hardening of the panel may not be what is expected.

The normal anisotropy of aluminium alloys is typically lower than that for steel (r value <1). The higher the r value the deeper a panel may be drawn and once again this indicates that product design must be consistent with the material that is selected. On the other hand, the lower the planar anisotropy, the more repeatable will be the forming behaviour under edge stretching conditions.

In view of the tendency for aluminium to build up in the draw die (which leads to scoring of the panels), it is recommended that an appropriate lubricant be selected with the aluminium supplier to minimize galling. It is common practice in North America to apply a compounded petroleum prelubricant at the rolling mill and later to apply a water-based lubricant in the blankwashers to 'clean' the prelubricated blanks of slivers. The combination of these two types of lubricant yields a coefficient of friction below 0.07.

According to the experience of automotive manufacturers in Europe, aluminium sheets with an EDT (electro-discharge technology) finish yield better forming performance than with the regular 'mill' finish. The EDT topography is such that lubricant is retained evenly across the entire surface of the blank. The optimum surface roughness which still guarantees a good paint appearance while avoiding fretting is considered to be $R_a = 1.0$ micron. It should be mentioned that although the requirements for surface quality may be more severe in Europe, one North American automaker adopted the EDT surface finish for a large volume hood outer panel but later resorted to an 'improved' mill finish. Mill finish may nevertheless be unacceptable for certain vertical applications where 'slumping' may occur (uneven surface developed on painting).

The shape and size of the blank are critical to the success of the drawing operation. It is recommended that developed blanks be used whenever this is possible as it generally leads to a more even flow of material during the drawing operations and consequently a more even strain distribution and a more predictable springback behaviour. The consequence of introducing a blanking operation is that a washing operation and a further application of lubricant will be necessary after blanking. It is also recommended that aluminium blanks be stored in a controlled environment (i.e. room temperature) for approximately 24 hours prior to being pressed.

As will be shown, there is a definite interdependence between the different forming operations which are required for the production of a typical outer body panel. In order to convey this in a graphic way the various parameters that influence the final quality of a hemmed subassembly have been summarized in the form of a chart, which is based on the forming limit diagram (FLD) (see Fig. 5.13). This chart

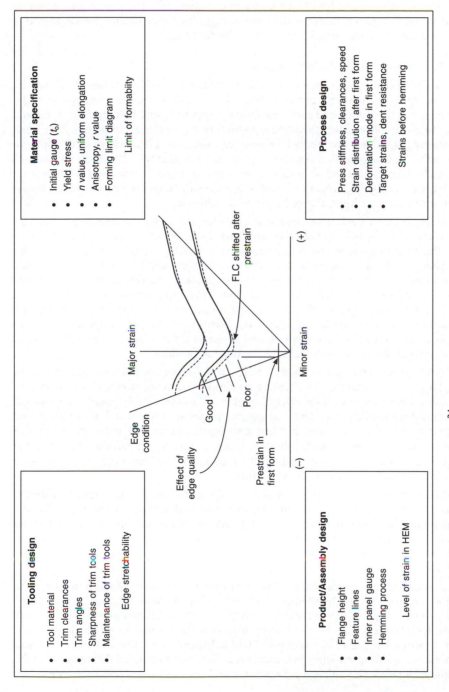

Tooling design

- Tool material
- Trim clearances
- Trim angles
- Sharpness of trim tools
- Maintenance of trim tools

  Edge stretchability

Material specification

- Initial gauge (t_0)
- Yield stress
- n value, uniform elongation
- Anisotropy, r value
- Forming limit diagram

  Limit of formability

Process design

- Press stiffness, clearances, speed
- Strain distribution after first form
- Deformation mode in first form
- Target strains, dent resistance

  Strains before hemming

Product/Assembly design

- Flange height
- Feature lines
- Inner panel gauge
- Hemming process

  Level of strain in HEM

Edge condition

Major strain

Effect of edge quality

Good

Poor

Prestrain in first form

FLC shifted after prestrain

Minor strain

(−)

(+)

Fig. 5.13 FLD related to hemming aluminium[24]

illustrates qualitatively the causes for loss of formability in an area of an outer body panel that is to be hemmed. The first parameter that affects the formability of aluminium panels is the material itself. Alloy composition, microstructure and crystallographic texture as well as material thickness have a direct influence on the shape and position of the forming limit curve (FLC) in the major/minor strain space.

Figure 5.13 also presents a line on the FLD, which corresponds with the mode of deformation at the free edge of a panel when it is stretched. The plastic anisotropy of the material (r value) determines the ratio of major to minor strains (the slope of this line) and therefore the forming severity at the free edge.

5.2.1.2 Product and process design

The design of outer body panels is largely dictated by the stylist's creativity. However, it is becoming increasingly common for 2–D and 3–D computer simulations to be carried out at the styling and early feasibility stages to ensure that specific styles and curvatures are indeed manufacturable. Assuming that material properties have been accurately determined the product designer can also take advantage of CAE software to assist in the design of sheet metal components (with flanges, rebate and addendum) that are both functional and manufacturable. The level of strain in a panel is determined to a large extent by its geometric features: draw depth, height of embossments, cross-section of character lines, flange height and radii, etc.

Beginning from the CAD data the process engineer will design the stamping operations that will enable a quality panel to be produced. Typically panels are produced in four or at the most five operations: a first draw and a possible redraw, trimming, piercing and flanging. In the case of a skin panel the shape, stability and dent resistance requirements are such that the minor strain throughout the central part of the panel should be between 1 per cent and 2 per cent. This requirement for a certain strain level in the panel signifies that the draw die must be developed accordingly. The addendum geometry, the binder shape, the position of the punch opening line, the height of any draw bars, the position and shape of draw beads and the blankholder force must all be designed to achieve the desired product.

Although subsequent operations will have an influence on the final quality of the panel it was found that the first forming operation (a draw or stretch-draw operation) has the greatest impact on the overall quality of an outer body panel: dimensional accuracy hemmability and dent resistance. It appears then that both product and process engineering are interdependent and together contribute to the production of a panel with a certain strain distribution. It also seems clear from this study that the most cost-effective way to design both the product and the manufacturing process is with the assistance of computer simulation.

Furthermore, the level of prestrain and the mode of deformation will vary in different areas of the drawn panel and thus the ductility will remain in the metal for the flanging and hemming operations. The dependence of formability on the prior strain history translates into a change of shape and position of the FLC. Figure 5.13 illustrates how the FLC can shift after the first forming operation and therefore indicates how product and process design are critical for the hemming operation.

It has already been mentioned that the most important feature of the first form is to exhibit a uniform strain distribution while achieving the required target strains at the centre of the panel. In order to obtain the required strains in an aluminium outer body panel, lock beads should be used rather than draw beads whenever this is possible. High strains at the centre of a skin panel will enhance its dent resistance but may also increase the splitting tendency in critical areas. It is recommended that material in the middle of a panel be stretched between 1 and 2 per cent strain and those areas that lie just within the trim line exhibit strains of 2–4 per cent. If these areas of the panel are stretched beyond 4 per cent strain the remaining ductility may be insufficient to allow for flanging and hemming operations.

Because of the galling and scoring tendency with aluminium panels, it is recommended that the tools in the draw die be constructed from cast steel rather than cast iron, which is porous. In order to further reduce galling and tool wear, tools should be chrome plated as this reduces the coefficient of friction without incurring unreasonable costs.

Many outer body panels display a feature or character line which enhances the cosmetic appearance of the vehicle. In order to avoid splitting the material at the radii of the feature line, it is suggested that the average strain across the feature line should not exceed 50 per cent of FLD 0 (the lowest point on the FLD). The average strain is calculated by comparing the actual length of the line through the feature with the shortest distance between tangency points.

5.2.1.3 Trimming operation

Following the first form a trimming operation separates the scrap from the product. However, aluminium being a relatively soft material the trimming operation can lead to a poor edge condition if the tooling is not constructed, adjusted and maintained adequately. If the edge condition is not sufficiently good (excessive burr height, presence of slivers, etc.) edge splitting is likely to occur during a stretch-flange/ hemming operation. In this case the mode of deformation along the free edge is approximately one of uniaxial tension and Fig. 5.13 indicates how the ductility necessary for edge stretching is dependent on edge quality. Furthermore the trim line development will determine the flange heights on the product and the tendency for edge splitting may be accentuated by an incorrect flange height.

Therefore, the tool design for the trimming operation has a significant influence on the success of the hemming operation that follows. The accuracy of the press, the type of material used for the cutting edges, the clearances between the punch steel and the die steel, the sharpness of the cutting edges, the amount of entry of the punch steel below the straight land are all factors that will influence the quality of the sheared edge. However, there is still a lack of knowledge relative to the design and maintenance of tooling for the production of aluminium panels.

The first recommendation regarding the trimming operation is to use the most accurate and up-to-date presses. It is important that the press be properly adjusted so that the ram remains parallel with the press bed and that it comes down in a consistent and repeatable way at every stroke. This is one of the best measures to guarantee a precise and consistent edge condition. Large, rigid pilot pins may also be used to guide the upper trim die.

The cutting edges should be constructed from good quality cast steel sections that are through hardened. The cutting edges must also be kept sharp and maintenance of the edges should be done by NC machining for greater accuracy. The clearance between punch and die steels should be from 6 per cent to 9 per cent of initial material thickness. Excessive clearance will tend to increase the amount of fractured surface and a rough edge will be produced. Insufficient clearance, on the other hand, will increase the amount of sheared surface. Frequent maintenance of the cutting edges as well as setting the correct tool clearances will minimize the burr height and improve edge-stretching behaviour during the flanging and hemming operation.

5.2.1.4 Flanging and hemming

Prior to subassembly, the edges of a panel are flanged down 90 per cent. This operation leads to a bending mode of deformation in material that has previously been drawn or stretched. The maximum surface strain along the bend line will be determined by the flange radius as well as the previous strain history.

A configuration study involved a luxury vehicle in which an aluminium door outer panel is hemmed to a zinc coated steel door inner panel of thinner gauge (with sealer separation to prevent bi-metallic corrosion effects). There are many types of hemming assemblies used in the industry featuring varying thickness of inner and outer panels and types of design. Figure 5.14 presents some common hemming designs: namely the rope hem and the flat hem. To avoid surface roughening or, worse, cracking, care must be taken in selection of a radius around half the thickness of the sheet, and some form of roping or local easing of the radius is normally required, e.g. by 'backing off' the flattening tool. Roping is a solution but again care is necessary in tool design at feature lines or corners to prevent wrinkling or related effects. However, one of the primary concerns with aluminium outer body panels is the ability of the material to be hemmed after having been work hardened during previous forming operations. The flat hem that is commonly used with steel outer panels tends to be too severe for most aluminium alloys and, again, therefore the relieved flat hem or the rope hem is recommended.[25] As the inner radius of the outer panel increases, the bending strains decrease but the box line of tile joint loses its crisp appearance. Therefore a compromise must be found between aesthetic appearance and overdesigning the hemmed joint.

It also appears that the bending behaviour of aluminium alloys greatly depends on the microstructure of the alloy. The FLD, for instance, is not suitable for predicting the bendability of sheet metal. A hemming study, which was carried out as part of the collaborative programme, indicated that certain aluminium alloys can be hemmed after a 5 per cent prestrain under plane strain conditions. First, this underlines the importance of designing the strain distribution in the first forming operation in such a way that the panel is not prestrained beyond a certain limit (e.g. 5 per cent) in the areas that are to be hemmed. Second, it must be emphasized that the mode of deformation during the prestrain is just as important as the level of prestrain attained in this first operation. Therefore the appearance of a hemmed subassembly may be unacceptable in an area that was subjected to 5 per cent prestrain in equibiaxial tension whereas it may be acceptable if the same prestrain were in a mode of plane strain. This consideration points again to the importance of product and process design.

Again, an aluminium alloy may perform well in the drawing operation but have a microstructure that is not optimized for bending. Once more, involving the material supplier at the early stage of design will enable the automotive manufacturer to select not only the right alloy but also the right production process for this alloy.

Moreover, it is recommended that a relieved flat hem be used with aluminium outer body panels particularly when the inner panel gauge is less than that of the outer panel. This particular design (Fig. 5.14) is a good compromise between aesthetic appearance with a fairly crisp box line but without leading to the severe bending strains that are induced by the flat hem. It is also recommended that the inner and outer panels be positioned in an accurate and consistent way and that the operation is carried out in a hemming press rather than a table top hemming.

5.2.1.5 Paint process implications for aluminium

During a paint shop audit, the actual time/temperature cycles of the baking ovens were recorded and related to responses predicted from simulated laboratory studies. In these studies the bake-hardening response was determined relative to the amount of cold work and heat treatments. In view of the different bake-hardening responses of 5xxx and 6xxx series alloys, different ideal cycles would be suggested relative to each alloy. In the event that 5xxx series alloys are used for outer body panels, the

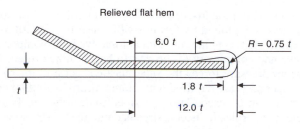

Relieved flat hem

Fig. 5.14 Recommended geometry for hemming of aluminium outer panels

ideal cycle of the electrocoat oven would be 160°C for 20 minutes. This would promote good paint adhesion while preventing excessive softening. If bake-hardenable 6xxx series alloys are used for outer panels, the electrocoat curing temperature should be higher. If optimum curing conditions are applied, up to 40 per cent increase of the yield stress can be achieved depending on the level of prestrain. (The optimum age-hardening response is achieved by a still higher heat treatment, e.g. 200°C for 60 minutes, which is normally not attained by electrocoat curing.)

Although the highest temperatures are attained in the electrocoat oven it would appear that subsequent surface and top coat cycles also contribute to the bake hardening of these panels. A decision therefore has to be made relative to the complete paint-baking cycle and the choice of alloy, but often a compromise must be accepted.

Surface texture

Traditionally, mill-finish sheet with a unidirectional texture impressed by machined work rolls has been used for body panels and while used for utility type vehicles provides an adequate surface for bonding and painting. As quality expectations have improved with regard to the external finish requirements for premium vehicles so the directionality gradually began to have a more noticeable effect, causing slight slumping of the paint when mounted vertically. The material is therefore currently specified with a 1.0 micron EDT finish and for maximum corrosion resistance a Zr/Ti pretreatment is preferred.

5.2.2 Superplastic forming

Superplastic forming (SPF) is a manufacturing technique using air pressure and single surface tooling. When heated within their plastic range (temperatures typically are 500°C), aluminium SPF alloys (Table 5.3) can be made to stretch to many times their original length. For an alloy to be superplastic it must have a uniform grain size,

Table 5.3 Typical room temperature properties of superplastic forming alloys

BS & International STD	ISO designation	Condition	0.2% proof stress MPa	UTS MPa	Elongation %
5251	AlMg 2	0	60	180	18
5083 SPF*	AlMg 4.5 Mn	0	150	300	20
5754	AlMG 3	0	100	205	25
2004 SPF*	—	0	130	220	9
2004 SPF*	—	T6	300	420	9
6061	AlMg 1 SiCu	T6	240	295	7

* Some alloys are defined as SPF alloys while some are superplastic versions of current alloy types.

typically less than 20 microns. This property of high tensile ductility at low strain rates and elevated temperatures is exploited by forming SPF sheet over a tool using air pressure. There are three forming techniques used for automotive body components depending on the size, shape and complexity of the part to be manufacturered, Fig. 5.15. In drape forming, the air pressure forces the superplastic sheet into the cavity and over the male tool. If deeper parts are required with more uniform wall thickness, then additional die movement can be used. This is termed male forming. Female forming is the process whereby superplastic sheet is stretched into the tool by applying air pressure.

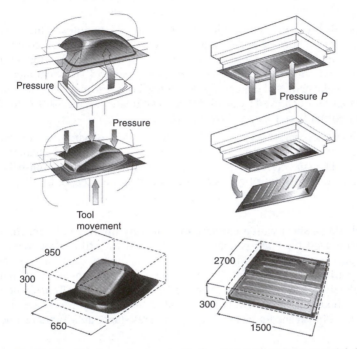

Fig. 5.15 Versions of superplastic forming (indicating potential dimensions achievable)

Advantages of the process include:

- Ability to form complex shapes with multiple curvature from a single sheet, thereby reducing component numbers and subsequent fabrication and assembly costs
- Short lead times for rapid prototyping
- Low cost single surface tooling for low production volumes
- Easy to achieve A class finish since only one surface of the tool is in contact with the sheet
- Limited spring back of panels

The main disadvantage of the process is the long time required to form a part, typically many minutes. This is because the superplastic forming process requires low strain rates. As a result, the process is only suited to low volume applications typically up to 10 000 parts.

Previous experiences in the field of automotive body structures have demonstrated the feasibility of this type of forming. Used in conjunction with low cost refractory

Fig. 5.16 Mini body panels produced from 'Prestal' superplastic sheet and rig showing refractory tooling used for forming at elevated temperatures

tooling and development studies carried out in 1970 resulted in a body shell for the classic Mini. Using refractory tooling and vacuum or pressure forming it was proved that this technology could be scaled up to produce a complete set of body panels and conventional spot welding and finishing applied to manufacture the complete car body. The vehicle was subsequently used as a factory runabout for many years.

Disadvantages experienced with early prototype parts made from the 'Prestal' Zn–Al alloy[30] included roping or longitudinal roughening of the surface (requiring rework of external panels) and tendency for creep at summer temperatures, although most of the advantages associated with the forming of plastics, including re-entrant shapes, were proven. With further alloy development this technology is still worth consideration particularly if lightweight versions of current models were rapidly required, e.g. for export markets.

5.3 Manufacture of components in magnesium

The major process for the manufacture of autobody components from magnesium is die casting. The cost of producing a die cast mould is generally less than for press tooling, making the application of cast components particularly suitable for lower volume applications. In pressure die casting the molten magnesium is injected into the mould cavity under pressure. As soon as the part is solid the mould is opened and the part ejected. Cooling water and mould release agent are then applied to the mould before the cycle proceeds onto the next part. Tool life can be of the order of 100 000 parts. Similar technology is applicable for both magnesium and aluminium although shrinkage rates will be different. A magnesium die casting was selected by Mercedes-Benz for structural use on its SLK model for the wall between the fuel tank and luggage compartment. This weighed 7.04 lb compared to 13.2 lb for a steel part and the choice was made because the large surface of the part could be divided into zones of varying thickness and strength, the more heavily stressed parts were up to 6.5 mm thick while the central zones were only 1.8 mm thick.[28] As stated, this probably represents the way forward regarding body applications and other companies[27] are now following the Fiat example of using a single piece magnesium casting for the crossbeam under the dashboard which replaces an 18-part spot welded assembly (Fig. 5.17).

The pressure die casting production process necessitated the use of an open section member (a) in place of a fabricated box made by spot welding. The resulting multi-web reticular section (b) achieved a 50 per cent weight reduction, 80 per cent increase in bending stiffness and 50 per cent increase in torsional stiffness.

Sheet forming of magnesium at room temperature is limited due to the hexagonal crystal structure of the material, and twinning characteristics allow only low levels of deformation. It is possible to increase the elongation considerably at elevated temperatures and at 235°C it has been shown that elongations of over 35 per cent have been recorded.[26] Forging has been proposed as an alternative means of component production but even near net shape techniques like precision forging are not really relevant to many body applications.

5.4 Production of polymer parts

The various types of polymers and general production methods have been described in Chapter 3, and now the production of components are considered. This section is based on extracts from the excellent publication of Hodkinson and Fenton[27] and

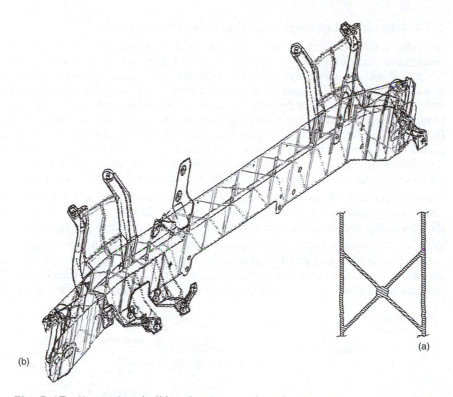

Fig. 5.17 Magnesium bulkhead cross member showing main casting and multi-web section[27]

includes reference to the RIM and RTM (resin transfer moulding) processes, and essentials of advanced GRP and SMC component and assembly manufacture. The technique used for the construction of Formula 1 racing car bodies was described in Chapter 4.

5.4.1 Plastic moulding for open canopy shells

The RIM process used in Bayer's metal/plastic composite construction also has important potential for production of structural panels for car bodies on a relatively high volume basis. In a punt-type vehicle structure, made from metal box sections stabilized by plastic cores and incorporating rollover hoopframes at the 'A' and 'C' posts, open shell sections in RIM polyurethane could be used to form the roof panel, and front/rear ends of the body superstructure enclosing windshield and backlight screen respectively. The punt-type structure also allows the possibility for a 'pillarless' sedan configuration with side doors hung from 'A' and 'C' posts, without the need for a 'B' post. Use of metal/plastic composite doors in conjunction with a structurally efficient sliding bolt system that would preserve the integrity of the door side-impact beams would allow unrivalled occupant access to the sedan interior. Open shells in RIM polyurethane could also be used for vehicle front and rear-end structures which would be 'canopies' suspended over purpose-designed shock-absorber systems cantilevered from the main punt structure to absorb front, rear and 'three-quarter' impacts.

5.4.1.1 Reaction injection moulding (RIM) developments

Involving polymerization in the mould, the RIM technique is quite different from other plastic moulding methods and can be used for producing quite complex parts and panels without undue high tooling investment – since mould pressure is low. Two or more components flow into the mixing chamber, at relatively high pressure (100–200 bar) and are then expanded into the mould at much lower pressure. The streams impinge at high speed to obtain thorough mixing and initiate polymerization as they flow into the mould cavity at a pressure of about 100 bar. Low viscosity during mould filling is one of the key attractions of the process as a relatively small metering machine can make large parts. The low viscosity also simplifies reinforcement with, for example, the possibility of using continuous-fibre mat placed in the mould.

Some 90 per cent of RIM production is in polyurethanes and urea-urethanes, the latter being uniquely suited to the process as they do not melt flow like normal thermoplastics and therefore conventional injection moulding is not possible.

ICS have developed a family of polyureas for body panel applications with unusually good processability and physical properties, Fig. 5.18. Gel times of 2 seconds are possible and mould temperatures are less than 93.5°C. Overall cycle time is about 1.5 minutes and further development promises 'less than 1 minute', with equipment

	RTM	S-RIM
Equipment cost	$30 000	$500 000
Flow rate (Kg/min)	2.3	55
Mixing	static mixers	impingement
Mould pressure (MPa)	0.3	2.4
Void content (vol%)	0.1–0.5	0.5–2.0
Mould materials	epoxy	steel
Mould temperature[b] (°C)	25–40	95
Component viscosities (MPa's)	100–550	<200
Cycle time (min)	10–60	2–6

(d)

Typical properties	
Specific gravity	1.10
Flexural modulus, psi	
at 73°F	144 000
at −20°F	220 000
at 158°F	95 000
Modulus ratio −20°F/158°F	2.31
Tensile strength, psi	5100
Elongation at break, %	95
Gardner impact, ft-lb at −20°F	10.1
Mould temperature, °F	160
Component temperature, °F	110

(b)

	Isocyanurate	Urethane	Acrylamate	Epoxy
Random glass mat (wt%) (3 mm thick part)	38	44.8	40	40
Specific gravity	1.54	1.53	1.46	–
Void (vol%)	1.5	1.5	–	–
E_f (MPa at 25°C)	8100	9600	8700	9200
Tensile strength (MPa)	150	150	125	160
Elongation (%)	7.3	2.0	2.1	1.2
Izod impact (J/m)	510	660	790	−800
Heat distortion (°C)	184	189	240	>200
Thermal expansion (m/m°C) × 10^{-6}	−20		27	18

(e)

Fig. 5.18 RIM process and properties: (a) RIM machine; (b) body-panel RIM formulation; (c) steps in the S-RIM process; (d) S-RIM and RTM compared; (e) S-RIM-composite properties

shown at (a). Filler packages are also becoming available which allow part surface finish comparable with steel; moisture stability is high compared with competing thermoplastics and the materials can tolerate temperatures of 190.5°C. The table at (b) shows typical properties for a formulation that would suit body panels but others are available which raise the elastic modulus as high as 200 000 psi.

As well as increasing strength and rigidity of panels, the addition of glass fibre or other reinforcements considerably improves the compatibility of thermal expansion coefficient with such materials as steel and aluminium. Since polymers typically have coefficients some ten times greater than steel, a metre-long part hung onto a steel body could change in length by 1 cm between summer and winter temperatures. S-RIM is the process, shown at (c), in which long-fibre mat is placed in the mould and reactive monomers injected onto it, the process being akin to resin-transfer moulding but with high pressure impingement mixing to accelerate the reaction. A comparison is made at (d) while typical properties of S-RIM composites are shown at (e).

5.4.1.2 Resin transfer moulding (RTM) to incorporate foam cores

For covers such as bonnet and boot lid, self-supporting horizontal panels can be made with either glass-fibre laminate or foam cores, effectively automating the sandwich panelmaking process but allowing complex shapes and variable thickness cores in one panel. Essentially low viscosity resin is injected into a mould containing the required preformed insert. For relatively short model runs (10–20 000), RTM, Fig. 5.19, is a lower capital cost process than SMC compression moulding. The stages of the process are shown schematically at (a). First the glass reinforcement or preform is placed in the mould. Once the mould is closed the resin is injected with no or little movement of the glass. After mould filling the part is left curing in the mould until it is dimensionally stable so that it can be demoulded without losing its shape. The fact that the reinforcement is pre-placed in the mould gives the process the potential for making parts with better surface and mechanical properties than SMC where the fibre orientation is usually less controlled and favourable because of its flow.

During the filling stage, the resin being injected usually contains filler and can exhibit a viscosity with some shear thinning behaviour. However, for typical RTM compositions this effect is small and as a first approximation can be neglected. According to the Seger & Hoffmann subsidiary of Dow Chemical, to increase the efficiency of the moulding stage it is best to perform as many operations as convenient outside the RTM mould; the optimal placement of fibre into the RTM mould can be time consuming particularly if the shape is relatively complex. Methods available for preforming complex shapes include spraying fibres and binders onto a perforated mould or the use of mats and/or fabrics pretreated with thermoplastic binder which can be shaped by pressure when heated.

A shell of a sports seat is one example successfully produced by the company. The part is complex with a deep draw, multi-curvature shape and multi-plane shut-off. The method of preforming chosen used continuous filament random glass-fibre mats. The preforming of fibre reinforcement for flat mouldings such as a bonnet may not be absolutely necessary. The handling of the fibre mats, however, can be difficult as they are rarely self-supporting. A bonnet can be designed as a slim sandwich structure instead of inner and outer mouldings. The rigid foam core thus provides a convenient method of supporting the fibre reinforcement during transportation of the fibre to the RTM mould, (b).

Epoxy resin systems, with their low volume shrinkage of between 1 and 3 per cent together with their good mechanical properties, are ideally suited to Class 'A' surface

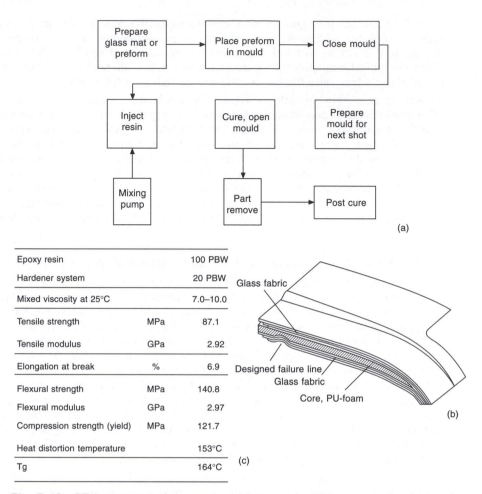

Fig. 5.19 RTM process and properties: (a) stages in RTM process; (b) designed failure line in bonnet panel; (c) properties of resin systems

body panel applications. Painting at temperatures up to 150°C can also be met with specifically formulated epoxy resin systems. The table at (c) gives the most important mechanical properties of a typical resin system developed for RTM. Faster curing systems based on vinyl ester resins can be used for RTM structural moulding where surface finish is less critical. Vinyl ester resin-based systems are capable of giving mould closed times of under 3.5 min. High speed mixing and dispensing technologies developed for the polyurethane industry are also being adapted for the RTM process.

Metal inserts to spread loads at high stress areas, such as hinge attachment points, can be incorporated into the RTM moulding. The shape of the polyurethane core moulding can provide a method of locating and transporting these inserts during the preform process. A particular design feature of the bonnet is the designed failure line. During high speed impacts the bonnet fails along this line.

RTM was successfully exploited by PSA in their Tulip concept battery-electric car, Fig. 5.20. The carmaker worked with Sotira Composites Group to develop the body structure which comprises just five basic elements bonded together. These five parts constitute both the exterior and interior of the vehicle bodywork, (a). In effect, the

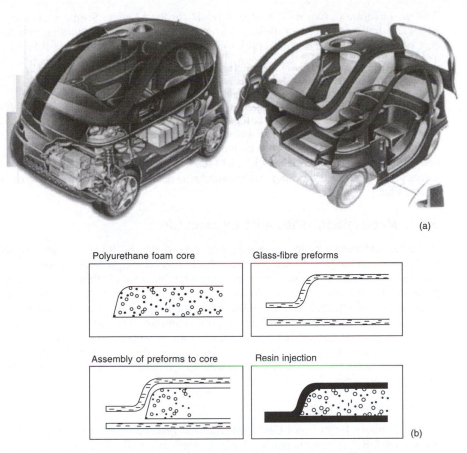

Polyurethane foam core

Glass-fibre preforms

Assembly of preforms to core

Resin injection

(a)

(b)

Fig. 5.20 Tulip concept car: (a) structure; (b) sandwich configuration

seats, dashboard, centre console and so on form an integral part of the structure which has significant benefits in terms of the rigidity of the vehicle.

Each of the five parts comprises a rigid polyurethane foam core of $110\,kg/m^3$ density. Glass-fibre mat is preformed and wrapped around the foam core before being placed into the low pressure injection tool. To ensure accurate location, the glass mat is retained in the part line of the tool. Polyester resin is then injected into the tool which impregnates the glass fibre and completes the sandwich construction, (b). With a 40 per cent by weight ratio of glass reinforcement to resin, the resultant assembly weighs aproximately 30 per cent less than an equivalent steel structure. The material used is also claimed to contribute to the safety of the vehicle, both for its occupants and to pedestrians, with energy absorbing characteristics of the panels shown to be 87 per cent higher than for standard steel parts.

To complement the precise resin injection system, the tools have a compression chamber in place of the traditional vents. The tools are also designed to maintain close temperature control across the entire surface. Typically $\pm 2°C$ is achievable to ensure consistent polymerization of the resin and the use of chromed steel or highly polished nickel shell tools allows for parts with Class 'A' surface finish to be moulded. Resin supplier DSM had a key role in the project to optimize the resin system to suit the RTM process. Mould flow analysis tests have therefore been performed with the aim of reaching body panel production rates of 200 per day.

The manufacture of Noryl GTX front fenders for the Mégane Scénic was described in 1997[29] by Debuigne, and illustrates some of the aspects of moulding and fixing this type of material. The GE GTX 914 material has been used for the R5 and Clio 16s but for the Scénic it was required that material formulation compatible with electrostatic painting was required. The GTX 974 variant does not contain fillers and therefore its expansion coefficient was ten times higher than that of steel, and the development of a special sliding fixture was necessary to accommodate this. Although not typical it was emphasized that all process aspects needed careful tuning of mould, drying system and screw of the injection machine. Another requirement was to obscure faults like the joint line, which could be accomplished on the Clio within the design of the upper fender. The equivalent feature on the Scénic was more visible and a special 'Lunette' split tooling arrangement was designed to change the line of sight of the joint.

5.4.2 Materials for specialist EV structures

Polyester and epoxy resins have a proven record for the lower volume specialist vehicle categories. The future electric vehicle market might tend on the one hand towards localized body manufacture with considerable manual labour glass-reinforced polyester (GRP) content in developing countries and in the richer countries to the construction of ultralightweight bodies using techniques thus far only affordable to race-car construction, Fig. 5.21. Hand lay-up is an important factor for both of these sectors. There is also a medium volume sector in specialist vehicles which has warmed towards resin pre-impregnated sheet moulding compounds which might well be adopted for electrical commercial and passenger service vehicles as the market progresses.

5.4.2.1 GRP and SMC

For particularly lightweight GRP construction, reservoir moulding is a method of producing GRP sandwich panels. Here a reservoir of flexible open-cell foam is impregnated with a resin and sandwiched between two layers of dry fibre reinforcement. The three layer sandwich is placed between dies under a pressure of just 12 kg/cm^2. The foam acts as a sponge which, during moulding, squeezes the resin into the fibre. Sandwich stiffness can be altered by varying compression pressure and the tooling can be simplified by using one rigid and one flexible die face such as a liquid-filled bag. The VARI (vacuum assisted resin injection) process was pioneered by Lotus for car body shells but is now available for a variety of licensed manufacture applications. Here the part is made between matched mould surfaces after laying up of the reinforcement by hand. Vacuum is then applied to the space between the mould faces and resin automatically drawn into the cavity. Preformed foam cores can also be placed in the mould to achieve localized box sections within the main part. It is also now possible to preform the glass reinforcement to speed the lay-up process. Another development is a technique for making metal-faced moulds which can be heated to further shorten the cure cycle. The view at (a) shows the bottom half of a Lotus car structure, made by VARI, with numbered panels indicating the weight of glass (in lb/ft^2) used in each area of the moulding.

The ability to produce sheets of glass-fibre reinforcement impregnated with a catalysed layer of polyester, within protective layers of polyethylene film, has, of course, led to sheet moulding compounds (SMC). These simplify the process of matched die moulding to give relatively high production rates. Low profile resin systems can be employed for exterior body skin panels requiring high surface smoothness. Four types of SMC reinforcement are common: short chopped glass rovings in a random

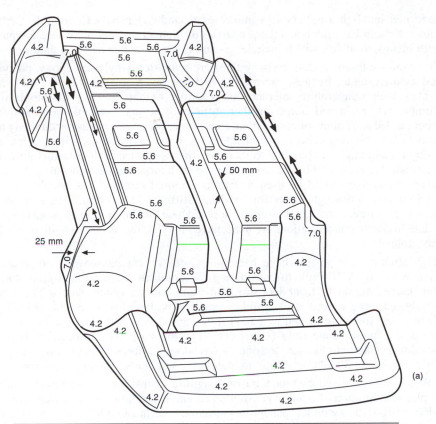

Fig. 5.21   Reinforced plastics: (a) weight of glass in moulding; (b) high strength composite properties

	Tensile strength (GPa)	Tensile modulus (GPa)	Specific gravity	Specific strength	Specific modulus
A-S/epoxy	1.59	113	1.5	1.06	75
XA-S/epoxy	1.90	128	1.5	1.27	85
HM-S/epoxy	1.65	190	1.6	1.03	119
S-glass/epoxy	1.79	55	7.0	0.90	27
E-glass/epoxy	1.0	82	2.0	0.50	21
Aramid/epoxy	1.29	83	1.39	1.00	60
Steel	1.0	210	7.8	0.13	27
Aluminium L65	0.47	76	2.8	0.17	26
Titanium DTD 5173	0.96	110	4.5	0.21	25

pattern for equal strength in all directions; endless rovings arranged parallel to chopped fibres for unidirectional strength; medium length rovings arranged parallel but staggered to obtain better mould flow than in the last case; and a wound formation of crossed roving tapes laid at 20° again to effect directional strength.

Structural performance comparable with metals can be had with the addition of materials such as para-aramids and carbon fibres. These materials are already commonplace in racing car structures and offer possibilities for such parts as slim-line windscreen pillars. The Kevlar brand of para-aramid, by Du Pont, exists in two states: first as a 0.12 mm thick endless fibre with cut length varying from 6 to 100 mm, and second in the form of a 'pulp'. It is said to be 30 per cent stronger than glass. The company also make a so-called meta-aramid, Nomex, which is a 'paper' that can be formed into lightweight honeycomb cores of sandwich panels. Courtaulds' Carbon Fibre Division have tabulated the properties of high strength composites, compared with metals, as at (b), which includes S- and E-glass composites as a datum.

The common thermosetting-resin systems are polyester, vinylester, epoxy, phenolic and bismalyamide. Because thermosets will not revert to their liquid state when heated, their temperature tolerance is generally higher than for thermoplastics. Compressing resin and fibres together sufficiently to get a reliable bond between them has led to systems of curing mouldings under pressure such as vacuum bag and autoclave. Designing in press-moulded GRP permits maximum parts integration into a single moulding; it is possible to increase stiffness by using double curvature and well-radiused curves (this also assists fibre distribution and air removal). As much taper as possible should be used to facilitate mould removal; use of ribs, hollow sections and lightweight cores for localized stiffness is recommended and use of large area washers and metal inserts to spread shear and bearing loads at attachment points (as stress concentrations are not relieved by yielding as with ductile metals) is recommended.

High strength composites have been proposed for car passenger compartment frameworks by VW. Prior to considering the complete framework, initial studies were carried out on the front door surround-frames; these, which showed 20 per cent weight reduction against steel, could be obtained without loss of rigidity, Fig. 5.22. Normal test forces subjected to a steel frame are shown at (a) and used to test the composite frame, made from 60 per cent (by weight) glass-reinforced SMC with two-thirds continuous filament and one-third random cut fibres. With carefully designed fibre orientation 30 GN/m^2 elastic modules can be obtained (against 210 for steel).

VW has also addressed the vital topic of designing against the fundamental weakness of plastics: duration of loading causing creep and relaxation. The research analysed GRP–PUR–GRP sandwich panels in particular and the graph at (b) shows creep strain of GRP while (c) shows creep of rigid PUR under constant shear. The graph at (d) shows time and temperature dependency of GRP elastic modulus, deformation being shown against a base of loading period. Maximum normal stress in the skins can, of course, only be reached if shear stress in the core, and its adhesive bond to the skin, is not exceeded. In the case of profiled parts, the researcher points out that the skin's own bending stiffness is no longer negligible so shear stress in the core is reduced and allows higher short-time structural loading. Under all load conditions the profiled structure shows better time-dependent mechanical behaviour than the flat one. VW research has demonstrated that despite the familiar creep of plastics under sustained load, the materials possess good recovery capacities due, the author says, to their exponential and linear division of creep functions. FEM analysis of a composite car body floor (e) with 1.5 mm GRP skin and 12 mm thick rigid PUR foam core was carried out to examine these load cases: temperature-dependent creep under bending by four 100 kg passengers; static simulation of front impact at 14 g with and without 20 g seat belt forces and dynamic loading induced by the sub-idle engine shake mode. The analysis showed that compared with a steel floor the composite panel showed higher stiffness even under the constant load case of 1.5 days at 60°C.

5.5 Learning points from Chapter 5

1. Quality and consistency of product is essential to maintain the high productivity of modern press lines, as due to the overall momentum of the system sudden failures can result in massive scrap or rework costs.

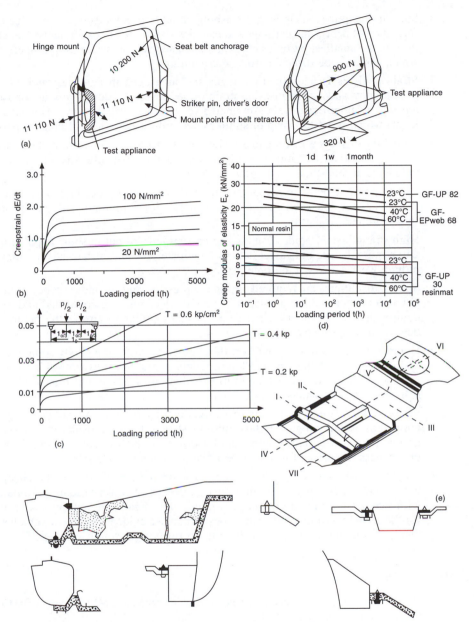

Fig. 5.22 Composites in passenger cars: (a) test forces on front door frame; (b) creep strain of GRP showing exponential and linear regions; (c) creep of rigid PUR foam under constant shear; (d) time/temperature dependency of GRP; (e) composite GRP floor design

2. A detailed specification should be written for each part based on predevelopment observations and past experience for similar parts, and regular monitoring of properties carried out to ensure consistently appropriate material is supplied for that part.

3. Strip properties should be monitored on a regular basis at the production site and press shop to ensure that the specification is met and, if necessary, batches can be presented to the press in 'smoothed' order. Presentation in order of ascending or descending values minimizes time lost on line stoppages to reset the press.

4. Special disciplines such as coil washing should be employed to minimize the formation of debris and particles associated with the use of coated steels. Transfer of even the smallest particles onto the punch surface can cause pimpling effects which can only be detected after final painting.

5. Strain analysis is a useful tool at die try-out stages in detecting and easing trouble spots and also in ensuring strain values developed during production runs are kept at subcritical levels. Automated systems are available for 3-D measurements which provide a map of strain ratios in different zones of pressed parts.

6. Modelling of sheet is starting to provide useful information on forming feasibility of new materials or tooling configurations but realistically this can only be indicative at this stage of programme development.

7. Due to improvements in manufacturing and processing procedures zinc coated steels can now be adopted for body panels with a minimum of quality and 'process chain' issues.

8. High strength steel grades can also be accommodated but in some applications will require the use of heavier duty tooling materials, and surface heat treatment in high wear areas, e.g. chrome plating.

9. Special disciplines are required during the pressing and secondary processing of aluminium sheet. Sliver poduction from cut edges must be controlled and rigid control of heat treatment carried out by the supplier to ensure the hemming behaviour is optimized.

10. Hydroforming is a fundamentally different method of forming body components and variable properties can develop in tubular parts. Sheet hydroforming is one method of ensuring a more uniform distribution of strain and improved dent resistance in skin panels.

11. Superplastic forming vastly improves the ductility of some materials at higher temperatures allowing complex/multiple shapes to be considered, although production rates currently allow application to only low volume models and unevenness of surface finish can be a feature of some materials.

12. Polymer panels can be produced using a number of methods but the cheaper tooling costs generally result in a cost advantage at lower volumes compared with steel or aluminium pressings. A more widespread solution must be found for the recycling of the numerous types of plastics evident in current vehicles if increased utilization is to be made in tomorrow's lightweight vehicles.

References

1. Davies, G.M., 'Pre-coated Steel: An Automotive Industry View', MBM International Coil Coaters Conference, London, 1993.
2. Butler, R.D. *et al.*, 'Sheet Steel for the Automotive Press Shop', *Special Report 79*, The Iron and Steel Institute, London, 1963.
3. Davies, G.M., 'Effects of Dimensional and Shape Deviation of Steel Products on Automotive Processing Operations', *Iron and Steelmaking*, Vol. 17, No. 6, 1990, pp. 381–383.
4. Pearce, R., *Sheet Metal Forming*, Kluwer Academic Publishers, Dordecht.
5. Davies, G. and Easterlow, R.A., 'Automotive Design and Material Selection', *Metals and Materials*, Jan. 1985, pp. 21–25.
6. Keeler, S.P., SAE Report No. 650535, 1965.
7. Goodwin, G.M., SAE Report No. 680093, 1968.
8. Brookes, I.C. and Davies, G.M., 'Gridmarking Techniques in the Press Shop', *Sheet Metal Industries*, Nov. 1972, pp. 707–710.

9. Sollac, *The Book of Steel*, Lavoisier Publishing, Paris.
10. Swift, H.W., 'Plastic Bending Under Tension', *Engineering*, Vol. 166, 1948, pp. 333–357, 357–359.
11. Hill, R., 'A Theory of the Plastic Bulging of a Metal Diaphragm by Lateral Pressure', *Phil. Mag.*, Vol. 41, 1950, pp. 1133–1142.
12. Zinkiewicz, O.C. *et al.*, *Numerical Analysis of Forming Processes*, John Wiley & Sons, 1984.
13. Wang, N.M. and Budiansky, B., 'Analysis of Sheet Metal Stamping by a Finite Element Method', *J. Appl., Mech., Trans. ASME(e)*, 1978, 73.
14. Tang, S.C. *et al.*, 'Sheet Metal Forming Modeling of Automobile Body Panels', 15th Biennial Congress IDDRG, Michigan, ASM International, 1988, pp. 185–193.
15. Butler, R.D. and Pope, R., *Sheet Metal Industries*, Sept.1967, pp. 579–597.
16. Davies, G.M. and Moore, G.G., 'Trends in the Design and Manufacture of Automotive Structures', *Sheet Metal Industries*, Aug. 1982, pp. 623–628.
17. Kopietz, V.J. *et al.*, 'The Deep Drawing Sheet Surface Condition in Manufacturing Automobile Bodies', *ATZ Automobiltechnische Zeitsschrift*, 100, 1998, 7/8, pp. 20–23.
18. Schey, J.A., 'Friction in Sheet Metalworking', Paper 970712, SAE SP-1221, 1997, pp. 87–112.
19. Bell, T., 'Surface Treatment of PM Components', *Materials and Design*, Vol. 13, No. 3, 1992, pp. 139–143.
20. Tertel, A., extracted from private communication to G.M. Davies.
21. Boyles, M.W. and Davies, G.M., 'Through Process Characterization of Steel for Hydroformed Body Structure Components', IBEC Paper 1999-01-3205, 1999.
22. Walia, S. *et al.*, 'The Engineering of a Body Structure with Hydroformed Components', IBEC Paper 1999-01-3181, 1999.
23. 'Aluminum for Automotive Body Sheet Panels', Publication AT 3, March 1996. The Aluminum Association Inc., Washington DC 20006, pp. 15 and 16.
24. Green, D.E. *et al.*, 'A Simultaneous Engineering Approach to Forming Aluminium Auto Body Outer Panels', IBEC'97 Conference, Stuttgart, Germany, 30 Sept.–2 Oct. 1997, pp. 131–136.
25. Wolff, N.P., 'Interrelation Between Part and Die Design for Aluminium Auto Body Panels', SAE Paper No. 780392, 1978.
26. Droder, K. and Jannsen, S., 'Forming of Magnesium Alloys – A Solution for Lightweight Construction', IBEC Paper 1999-01-3172, 1999.
27. Hodkinson, R. and Fenton, J., *Lightweight Electric/Hybrid Vehicle Design*, SAE International, Butterworth-Heinemann, Oxford, 2001.
28. Phelan, M., 'New Cars, New Materials', *Automotive Industries*, Sept. 1996, p. 54.
29. Debuigne, M., 'Noryl GTX Front Fenders of the Megane Scenic Technical Points and Industrial Challenge', IBEC '97 Proceedings, Stuttgart, Germany, *Advanced Body Concept and Development*, pp. 107–112.
30. North, D., 'Superplastic Alloy for Autobody Construction', British Deep Drawing Research Group Colloquium on Heat Treatment and Metallurgy in Metal Forming, Univ. of Aston, 20 March 1969, Paper No. 5.

6 Component assembly: materials joining technology

Objective: The techniques used for the assembly of panels into the body structure are described starting with reference to the spot and fusion welding methods traditionally used for steel, before progressing to modified procedures required for aluminium and use of adhesives and mechanical fastening methods often utilized in lightweight designs.

Content: Essentials of resistance welding of steel are outlined – basic variables are introduced and the concept of welding lobe for setting splash/no-weld boundaries is explained – effects of zinc coated steel types are considered – process modifications required for high strength steels are explained – major differences for welding of aluminium are summarized – principles of adhesive bonding are introduced together with low and high strength products – joint durability and the need for underlying pretreatment application are outlined – consideration is given to the increasing use of laser welding and mechanical fastening.

6.1 Introduction

The durability and structural performance of any autobody structure is largely dependent on the quality and the design of the component joints. For maximum performance the joint should be designed with parallel consideration given to the joining process, especially with respect to accessibility to the parts to be joined in the production environment.

There are three fundamental options for joining autobody materials and components:

- Welding
- Adhesive bonding
- Mechanical fastening

Each of these joining modes will be discussed in the following sections starting with mild steel and then coated and high strength grades, followed by aluminium. The joining of plastics is a specialized and complex subject, complicated by the wide number of polymers in use, even for automotive panels. Although the same basic methods of ultrasonic, laser, induction, friction and solvent welding can be used these lie within the province of the supplier, for most polymer parts are outsourced and arrive at the automotive plant as bought-in assemblies. If expert advice is required it is recommended that this is obtained from a specialist source such as The Welding Institute (www.twi.co.uk) in the UK or sister organizations worldwide, and the same advice applies to other specialist forms of metallic material, such as duplex sandwich material, mentioned in the text.

6.2 Welding

Since it is by far the most common materials joining process in the automotive industry, most attention will be given to resistance spot welding (RSW) of coated steels. However, the major characteristics, advantages and drawbacks associated with the other joining processes are described briefly from materials, design, production, and processing perspectives.

6.2.1 Resistance welding

Welding is the joining of two or more pieces of metal by the application of heat and sometimes pressure. Resistance welding is that area of welding whereby the heat is generated by the resistance of the parts to be welded to the flow of an electric current. The main difference between resistance welding and other welding processes is that no filler metal or fluxes are used. In addition, fusion welding processes do not require the application of force to forge the heated workpieces together.

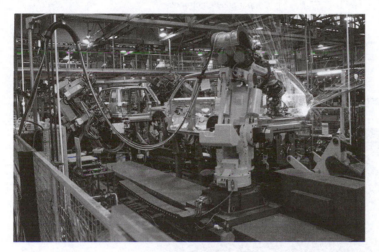

Fig. 6.1 Resistance spot welding of the BIW. (Picture courtesy of Land Rover)

However, even within the generic term 'resistance welding' there is a further classification of distinct processes:

- *Spot welding.* This is the most widely used example of a lap joining process, whereby the electrodes conduct the welding current and apply the welding force. The shaped electrodes (made from either a precipitationed strengthened copper–chromium and/or zirconium alloy or a dispersion strengthened copper alumina system) are held stationary when the weld is produced. Features of the resistance spot welding process can be seen in Fig. 6.2, and a typical cross-sectioned resistance spot weld in mild steel can be seen in Fig. 6.3.

- *Projection welding.* This is resistance welding in which the force and current to make the weld is localized by the use of a projection raised on one or more of the sheet surfaces. The projection collapses during welding. This process is commonly used for the attachment of mechanical fixings to the autobody structure, e.g. weld nuts.

- *Seam welding.* This is a continuous weld made by a single weld bead or a series of overlapping spot welds. The electrodes are revolving wheels and the continuous seam weld can be used to construct water-pressure-tight containers, e.g. petrol tanks. Interestingly, a version of the process was used for the attachment of the body structure bodyside to the roof on the classic Austin Mini.

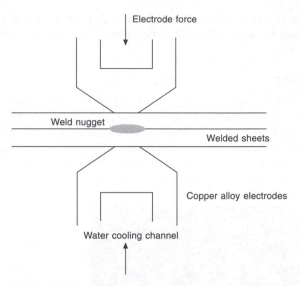

Fig. 6.2   Resistance spot welding process

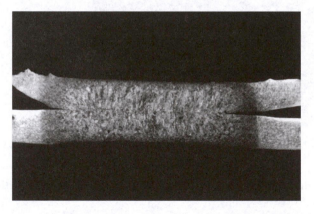

Fig. 6.3 Micrograph of a resistance spot weld in two sheets of mild steel

An important characteristic of the spot welding process is the short weld time used, thereby enabling high welding speeds. The time taken for production of a weld can be less than one second, depending on the specific application. In addition, the process is highly adaptable to automation and robotic techniques. A typical BIW contains approximately 5000 welds and the strength and durability of the automobile is largely dependent on the quality of resistance spot welding. The experience gained during the last 20 years with resistance spot welding coated steels means that consistently sound welds can be achieved at rates exceeding 20 million per week. Indeed, it is doubtful whether the modern car would be economically viable if it were not for resistance welding techniques.

6.2.1.1 Weldability

To fully appreciate the implications of joining technology on materials selection in the process chain, an understanding of the basic principles of resistance welding technology is required.

The main welding parameters in electrical resistance spot welding are: electrode force, welding current, welding time, squeeze time and hold time. For a basic spot weld, the relationship between the electrode force and welding current, as a function of time during the production of a weld, is shown in Fig. 6.4.

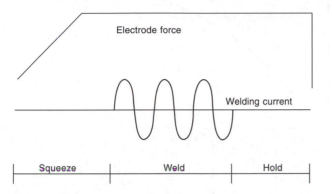

Fig. 6.4 Relationship between electrode force and weld current during production of a resistance spot weld

Thus, initially during the squeeze time, the electrodes clamp the sheets to be welded under the action of the electrode force, preset by the welding engineer. The squeeze time should be sufficient to overcome poor fit-up of the parts to be welded and in practice is similar for both uncoated and coated steels. Following the period of squeeze time, the welding current flows through the welding electrodes and the sheets to be welded. It is known that higher welding currents are required to weld zinc coated steels compared to the same thickness of uncoated steel. The weld time is defined as the time during which the current flows and when weld formation and growth occurs. Longer welding times are generally required to satisfactorily weld zinc coated steel compared to uncoated steel. The hold time is the time during which the electrode force is maintained, after the welding current flow ends, its purpose being to consolidate the weld. Too high an electrode force can result in expulsion or splash from the weld interface. Typical welding conditions are shown in Table 6.1 for coated and uncoated steels, though these will vary considerably depending on machine type.

Table 6.1 Typical spot welding conditions for coated and uncoated steels

Welding parameter	Coating type			
	Uncoated mild steel	Hot-dip zinc coated steel (GI)	Electrozinc coated steel (EZ)	Galvanneal steel (GA)
Welding force (kN)	2.5	2.8	2.8	2.7
Welding current (kA)	8	10	9.5	9
Weld time (cycles)	8	10	10	8

More detailed spot welding conditions for welding a particular material are best depicted in terms of a weldability lobe within which weld quality can be guaranteed, i.e. for a given weld time there exists a range of welding current that will produce acceptable welds. Typical limits used to produce a weldability lobe are an upper limit corresponding to interfacial splash and a lower limit corresponding to a minimum weld size, this depending on the product specification. A typical weldability lobe is shown in Fig. 6.5.

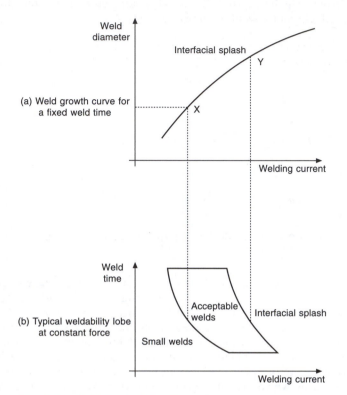

Fig. 6.5 Weldability lobe for resistance spot welding

For a fixed weld time, the size of the weld nugget will increase with current, according to the relationship shown in Fig. 6.5(a). At low current levels, a bond will form between the sheets without the formation of a weld nugget, this is termed the stuck weld condition. This should not be confused with the process of electrode sticking whereby a bond can form between the welding electrode and the sheet during the formation of a weld. At a particular current level X, the weld diameter will reach what is regarded as the minimum acceptable weld diameter, often defined as $3.5.\sqrt{t}$ where t is the single sheet thickness. As the current is increased beyond this level, the size of the nugget will increase until a stage is reached when the weld pool can break through the surface or be expelled from between the sheets. This is achieved at position Y and is termed expulsion or splash, a condition unacceptable irrespective of weld diameter. Therefore, for a given weld time there exists a current range that will produce acceptable welds. This type of exercise can be repeated for varying weld times and an acceptable range of current can be defined for each particular weld time. These combinations of current and time that produce acceptable welds are then expressed in the form of weldability lobes, Fig. 6.5(b).

The size of the lobe is thus a measure of the weldability of the material and suitable production welding configurations, with larger lobes indicating greater tolerances to changes in production conditions. The width of the weldability lobe is dependent on the basic resistance of the steel or coating being welded, which is a function of the thickness of the material, i.e. the weldability lobe width for 2 mm material is likely to be greater than 2000 A, whereas the lobe width for 0.6 mm material will be around 1000 A. Differences also occur between the weldability lobes for coated and uncoated steels, with the weldability lobe for a comparable coated steel shifted to the right on the x-axis in Fig. 6.5, i.e. to higher currents.

The increasing use of coated steels is of concern to the welding engineer because the presence of the coatings can, under certain circumstances, create difficulties when using resistance welding processes. It is also important to note that due to the demands of vehicle design described in Chapter 2, rarely are weld joints designed with the simplicity of two similar sheet thicknesses and it is common for three different thicknesses of sheet steel, with different coating combinations, to be welded in high volumes. Existing technology allows the production of a large number of welded components at high production rates for long periods. However, to maintain these production rates involves relatively high welding costs and necessitates a high level of shopfloor operator awareness, the latter being difficult to achieve over long periods.

6.2.1.2 Factors influencing electrode life

A distinction can be drawn between the ability to make a resistance spot weld, as defined by a weldability lobe, and the ability to continue to make a weld, i.e. the electrode life. The latter is one of the major difficulties encountered when welding coated steels. The electrode life can be defined as the number of welds that can be made before the weld diameter decreases below that previously specified and, in general, the electrode life is much lower for coated than uncoated steels. Typical electrode life values are given in Table 6.2.[2] In addition, the electrode life obtained under nominally identical welding conditions, can vary by over 100 per cent when welding zinc coated steels. Thus, for a hot-dip zinc coated steel while an electrode life of 1000 welds may be obtained in an electrode life test, a subsequent repeat test may produce an electrode life of 2000 welds. For manufacturing plants seeking a high degree of consistency in their operations, this situation may be problematic if good maintenance procedures are not enforced. While electrode life is affected by many factors, the most important factors which need to be considered include the electrode material and shape, the welding configuration (welding machine, current, electrode force) used and the material being welded.

Table 6.2 Typical electrode life values for autobody sheet materials

Material	Typical production line electrode life
Uncoated steel	>10 000
Electrogalvanized steel	<2000
Hot-dip galvanized steel	<1000
Galvanneal steel	2000–3000

The mechanisms by which failure of the welding electrodes occurs has been the focus of extensive research over the last few decades.[1,8] It is now generally agreed that the processes contributing to the failure when welding coated steels include:

- Softening of the electrode material due to the heat involved in the welding process
- Alloying of the zinc coating into the copper electrode causing the formation of brasses at the electrode tip
- Deformation of the electrode tip leading to an increase in the tip diameter and a decrease in current density
- Pitting and cratering of the electrode contact face

Before   After

Fig. 6.6 Electrode condition before welding and after a production shift welding coated steels

The exact contribution of these mechanisms is dependent on the coating type and results in the different electrode life exhibited. The typical appearance of welding electrodes before and after a production shift as a result of the electrode degradation mechanisms can be seen in Fig. 6.6.

The effect of coating thickness on electrode life is small compared to other factors. Research work has concluded that electrolytically deposited Zn and Fe–Zn alloy coatings are generally better than hot-dip equivalents.[3]

The implication of a low or inconsistent electrode life on production costs can be very significant. Production build schedules will be planned to allow electrode dressing to take place at predetermined intervals. Similarly, changing of electrodes usually takes place during a shift change. Failure of electrodes before these preset periods will result in unplanned downtime and loss of production. To avoid this, some form of short-term corrective action is necessary based on either electrode dressing or increasing the welding current using a stepping programme. Electrode dressing is based on the cutting or machining of the welding electrodes to remove brass alloy from the electrode tip and maintain the initial electrode tip diameter.

Current stepping is based on increasing the welding current at predetermined intervals to maintain the current density at the electrode/sheet contact face. Since the rate of tip growth varies when welding different coated steels so too will the required electrode dressing frequency/current stepping.

6.2.1.3 Weld testing

Spot weld testing can broadly be divided into those test methods used in production on the shopfloor for quality control purposes and those used for basic weldability confirmation of new materials and providing design data. When tested resistance spot welds can fail in one of three ways, Fig 6.7:

- Plug failure in which the crack propagates in the region around the weld leaving a slug or button of metal on one sheet
- Interface failure through the weld metal, along the original interface between the sheets
- Partial plug failure, a combination of plug and interfacial failure

It should be recognized that interface failure does not necessarily indicate brittle fracture, since close examination of the failure surface often reveals a ductile fracture surface.

The acceptance criterion for the chisel test is achieving a slug diameter of a given

Fig. 6.7 Spot weld failure modes

size dependent on the sheet thickness. Typically 3.5 to $5\sqrt{t}$ where t is the sheet thickness, this value depending on the manufacturers' specification or standard applied.

The second group of tests used for weldability confirmation of new materials and design data derivation include tensile shear, cross-tension, torsion, fatigue, impact and hardness and there are a number of European and international standards available to describe the correct test procedures for these.

The traditional method of spot weld quality control in manufacturing operations is based on a combination of non-destructive and destructive chisel testing. In non-destructive testing a tool is used to pry apart the welded sheets in an accessible position to assess whether a weld nugget has formed. This is often carried out at regular intervals during production. The destructive method comprises physically tearing apart the welded members to measure weld nugget size.

These methods may be considered to be wasteful and inefficient and the requirement has long been established for a suitable NDT technique that is cost effective and easy to use by only low skilled operators. Ultrasonic techniques may provide the answer and have already been introduced on a limited basis. Although a high rate of success is claimed in production situations,[4] the unavoidable requirement for a skilled operator is a major barrier to the widespread implementation of this technology. Efforts, therefore, need to be concentrated on developing a suitably simple NDT technique, most likely, though not necessarily based on ultrasonics, that the industry can embrace with confidence. This would provide a significant benefit to assessment of weld quality in HSS, sandwich steels and possibly multi-thickness weld configurations. At present such a system appears some way from development.

6.2.1.4 High strength steels

Despite some of the problems encountered when welding coated steel, e.g. electrode wear, the previously described processes are carried out successfully in BIW plants throughout the industry. Automotive companies are generally reluctant to move away from this technology. However, use of alternative material types has an influence on the effectiveness of the resistance welding process.

Virtually all commercially available high strength steels can be spot welded successfully. However, the key requirement for body assembly planning engineers is that similar equipment to that used for traditional mild steel grades can be used to weld high strength steels. In addition, welding conditions should be identical for different batches of material and the different suppliers of the same nominal high strength steel grade. Due to the difference in processing techniques within steel plants, the attainment of certain mechanical property levels may require differing steel chemistries. This may have a resultant effect on material resistivity and ultimately the welding conditions required to achieve a given weld size.

With certain restrictions on composition, welding parameters for high strength steels are not dissimilar to those for mild steel. Welding current may be slightly lower because the bulk resistivity of the steel will be higher due to the alloying additions within the high strength steel, e.g. C, Mn, etc. Certainly higher electrode forces will be required. Springback in HSS may also become an important issue in production resistance spot welding. Springback will require even higher than normal electrode forces to pull panels/components together. These higher electrode forces may deflect the welding gun arms and cause skidding of the electrodes on the surface of the components, with a resultant loss of weld quality. The requirement for panels with excellent dimensional tolerance will be met as steel suppliers ensure material grades are produced with consistent properties.

The main problem with high strength steel grades has historically been the occurrence of partial plug and interfacial failure types, as opposed to a full plug, Fig. 6.7. While welds failing in these modes can be of the same nominal strength, common production practice is based on acceptance of plug failure only, since this is an easy indication of weld quality on the shopfloor. When formulating the composition of new HSS grades, steel suppliers will attempt to concentrate on grades giving low C chemistries. High levels of Si, P, and C are avoided if acceptable weld properties are to be achieved using normal high speed welding conditions. This is indeed the case for new interstitial-free high strength grades.

The tensile shear strength of spot welds increases with increasing steel strength. However, the fatigue performance of spot welded joints is relatively independent of base steel strength, i.e. the failure of the spot weld joint with alternating stress is caused by the notch effect of the weld and not the steel strength. The advantages of higher strength steels can be used if one ensures the spot weld joints will not be stressed with critical strains. Alternatively spot weld pitch can be reduced and spot weld diameter can be increased to improve fatigue performance.

Stainless steel

Resistance spot welding of stainless steel has been confirmed as a viable process. However, a number of issues need to be addressed to achieve consistent weld quality. Increased electrode forces are generally necessary (up to 20 per cent higher than with mild steel). In terms of alloy content, there is some preference to reduce Ni content to less than 4 per cent to reduce costs and ensure there is no long-term health issues associated with the use of this material (the addition of Mn is made to maintain a suitable microstructure). Mixed joints consisting of stainless steel to zinc coated steel are achievable but with lower process stability. With a much higher resistance than mild steel, it is easy to weld very dissimilar panel thicknesses incorporating stainless steel, giving the added benefit of design flexibility in panel configuration.

Aluminium

All the automotive grades of aluminium sheet can be resistance spot welded and a number of low volume examples of resistance spot welded aluminium body structures have been produced including the Ford P2000, Fig. 6.8.

However, in comparison to coated steels there are certain differences to observe to optimize weld quality and productivity:

- Due to aluminium's low electrical and thermal resistivity, approximately two to three times the welding current and one-quarter the weld time are needed compared to steel. This will have implications when considering the cost implications of a spot welded aluminium structure, e.g. higher currents will require larger welding

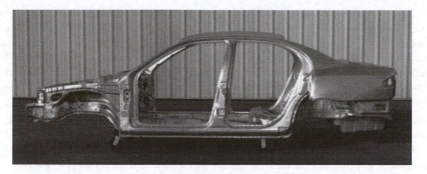

Fig. 6.8 Resistance spot welded aluminium body structure: Ford P2000

transformers and higher energy consumption resulting in additional manufacturing costs. Typical welding conditions for the welding of sheet aluminium in automotive applications are shown in Table 6.3.

Table 6.3 Typical welding conditions for sheet aluminium

Parameter	Typical aluminium conditions
Current	15 to 30 kA
Force	3 to 10 cycles
Weld time	2.5 to 4.5 kN

- As a result of the narrow plastic range (temperature difference between softening and melting temperatures) and the high thermal expansion, the accurate control and synchronization of current and electrode force is required to contain the molten weld nugget.

- Surface scale or any other type of impurity on the sheet surfaces can be incorporated into the weld causing weld inconsistencies and weakening of the weld joint. Suitable material preparation and cleanliness is essential to maximize weld quality.

- Due to the rapid rate of electrode degradation, a low electrode life will be encountered. This needs to be planned for when estimating production schedules.

In resistance welding there are a number of component resistances between the electrodes that govern the generation of heat, e.g. electrode to sheet contact resistance, sheet resistance, sheet to sheet resistance, etc. When welding steel the sheet resistance is sufficiently high to influence formation of the weld. For aluminium sheet, the low bulk sheet resistivity of aluminium means that the surface resistance of the aluminium sheets is extremely important in governing weld formation. For this reason, it is important that the surface properties of the sheets are consistent. Various mechanical or chemical treatments are recommended by sheet manufacturers to provide a consistent surface resistance. These include abrasion or etch cleaning. It is important that manufacturing scheduling is planned and accounts for the varying shelf-life of the treated sheets which can vary between a few days to a few weeks. These techniques can mean that increased currents are required to form a weld and subsequent stabilizing pretreatments are recommended to lower welding current requirements and extend electrode tip life.

Single-sided spot welding

It has already been established that the incorporation of hydroformed tubes into the vehicle structures, offering benefits in terms of improved sectional efficiency, may increase in future body structures. Fusion welding processes will be the most process tolerant joining technique for incorporating such parts, but with the disadvantage of potentially increased distortion after welding due to high levels of heat input. Laser welding would solve this problem but with the limitation of requiring parts of excellent dimensional accuracy and fit-up. Single-sided spot welding may be an alternative technology that finds limited application in production and there are reports of its use in the United States.

In situations where access is possible from both sides of a hydroformed component, a direct spot welding configuration can be used. However, with certain designs this will not be possible, e.g. a hydroformed sill to floor pan joint. In this instance, a single-sided welding set-up must be used with a series, indirect or remote welding configuration. With these configurations, the current return is via the steel component. The advantage with a series configuration is that two spot welds can be produced during each weld cycle. With indirect welding only one spot is made, the purpose of the flat electrode is to provide a current return path where the distance between the two electrodes is constant.

Welding conditions can be significantly affected by the section design, material thickness and mechanical properties of the components welded by this process. Thin and wide hydroformed components will require lower electrode force to reduce mechanical collapse of the section. Too low a force may result in void formation. The use of too high an electrode force can also produce hot collapse of the weld nugget into the hydroformed tube component. Crack formation is another common defect.

6.2.2 Fusion welding

Arc welding is used in autobody construction, especially where stress levels are considered to be high and confidence in a suitable resistance spot welded joint would be low. Historically, most arc welding is performed manually. However, productivity and the operating environment (fumes, heat generation) can be an issue and attention has turned to the use of robot applied arc welding. Unlike spot welding where there is a degree of flexibility on spot weld position on a sheet metal flange, in arc welding the weld has to be placed along a joint between pieces of metal. However, in a production BIW operation there are often part fit issues due to component tolerances, combining to form unfavourable minimum and maximum values or part distortion. These can mean accurate fixing of large sheet metal components is impractical by a robotic arc welding process. Nonetheless, recent applications are now in production, with research having focused on methods of accurate weld seam tracking.

Most applications of arc welding in the automotive industry are metal inert gas (MIG) welding with a consumable wire electrode. The arc is struck between the electrode and the workpiece and is shrouded by a layer of inert gas. The main advantages of the process are the good strength of the welded joints and the requirement for access from only one side. However, this must be offset by concerns over cycle times and in the case of aluminium, potential thermal distortion, shrinkage stresses and the requirement in some instances for surface pretreatment. An alternative for thin sheets (less than approximately 1.0 mm) is TIG welding.

The arc welding of thin walled aluminium extrusions (in relation to aluminium spaceframe construction) has its own unique problems, Fig. 6.9. Similar heat energy levels to those used for steel are required for aluminium. However, the lower melting point means that control of the molten weld pool, weld appearance and weld penetration can be problematic. There is potential for overpenetration as heat generation increases and material distortion occurs. If there is inadequate heat input underpenetration and poor weld appearance can result. Some weld defects can affect mechanical strength

Fig. 6.9 Arc welding of aluminium extrusions

under static loading conditions where they decrease the effective section of the weld. Internal defects such as porosity will not affect dynamic performance to such a large extent as geometrical defects like poor penetration or undercut. These issues appear to have been overcome on products such as the Audi A8, BMW Z8 and Honda NSX.

6.2.3 Laser welding

The use of laser welding as a materials joining technology has moved from a near reach technology to production application in the last two decades. Figure 6.10 shows a broad overview of laser processed parts which are cut or welded by CO_2 or Nd-YAG lasers in various automotive plants. In 1998 the ULSAB project specified 18 286 mm of laser welded joints in the final design.

The main benefits of the process are:

- High static and dynamic stiffness of the resultant joints
- The requirement for only one-sided access providing design flexibility
- The low thermal distortion of the joint
- The visual quality of the joint
- Reduced weight through the reduction of flange sizes
- Improved structural stiffness as a continuous joining technique

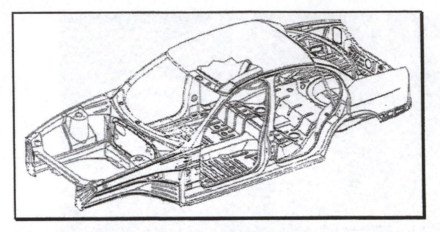

Fig. 6.10 Laser processed parts in the ULSAB BIW

However, the high investment and running costs still mean that the application of this technology is confined in the main to premium priced vehicles, e.g. BMW 5 series, Audi A2.

There are a number of considerations when selecting material types with respect to laser welding. A problem is associated with the low boiling point of zinc which can limit the application of laser welding of lap joints with zinc coated steels. The zinc coating at the interface of the two sheets (if double-sided coated) vaporizes during the welding process. With the sheets clamped firmly together with no gap present, the vapour must escape through the weld pool. This can cause expulsion of weld metal and any remaining vapour will be present in the final weld in the form of porosity. This porosity may be evident on the surface of the weld as roughness and surface pores. Techniques are developing to overcome this inherent problem either through new beam technology or technologies to maintain a controlled gap at the interface to allow the zinc vapour to escape.

Laser welding of aluminium for autobody applications is a newer technology. In addition to the general disadvantages of laser welding steel, such as high initial investment cost and the requirement for highly dimensionally accurate components, the laser welding of aluminium is also hindered by some specific material issues. The high thermal conductivity and reflectivity of aluminium mean that a high output laser and narrowly focused beam are required. The required power for welding aluminium with CO_2 lasers is greater than 5 kW and when using Nd-YAG is greater than 3 kW. Typical speeds are 5–8 m/min, although a maximum speed of around 12 m/min probably exists. The welding parameter range under which quality welds can be produced is somewhat narrower than that exhibited by steel. This requires even greater attention to process control.

It is interesting to note that unlike the other joining processes, the presence of the oxide film on the aluminium surface is actually beneficial to the laser welding process, since it lowers the power requirements for weld formation due to its better absorption properties. Close attention needs to be paid to component fit-up, sheet composition and process parameters to avoid difficulties with hot cracking and porosity. Edge preparation is important and gaps over 0.15 mm usually require a filler wire. Despite these difficulties, recent developments include the application of approximately 30 m of laser weld on the aluminium Audi A2. Most aluminium sheet manufacturers are investigating the use of laser welded aluminium blanks for cost reduction.

6.3 Adhesive bonding

Adhesive bonding technology has been used cost effectively for a number of years in a number of trim assembly applications to attach carpets, mirrors and for the attachment of stiffeners on the inside of doors, bonnets and hoods, etc. Attention is now increasingly being focused on the application of this technology to sheet component applications for load bearing joints. As a technology it is particularly suitable for application with thin coated steels and can be used in combination with mechanical fasteners or spot welding (weldbonding) to enhance the fatigue strength of connections in high strength steels. The advantages of adhesive bonding include:

- Ability to join dissimilar materials that are not weldable
- Provides uniform stress distribution, increasing structural stiffness
- Improvement in NVH performance
- Sealing of joints

Adhesive types are generally classified on the basis of the way in which the transition from liquid to solid takes place, i.e. either physical or chemical setting. Adhesives which set physically are called hot melt adhesives. Adhesives which set chemically include the curing adhesives. The two types of adhesives used to bond steel in automotive applications are thermosets and thermoplastics, with each offering different properties. Thermosets generally possess high shear strength, stiffness and durability. They are cured through the action of heat in the case of one-part adhesives or through the use of a curing agent in the case of two-part adhesives. Thermoplastics are particularly suited where energy absorption due to impact loading is likely. They are ductile, tough and perform well at low temperatures. Typical characteristics of structural adhesives are shown in Table 6.4.

Table 6.4 Common autobody adhesives

Type	Pros	Cons
PVC	Cheap, easy to apply	Overheating can produce corrosive by-products
Nitrile phenolic	Cheap, easy to apply	May retain moisture
Rubber based	Good anti-flutter properties	Cannot meet higher structural strength requirements
Epoxies	Capable of meeting higher strength requirements	Health and safety requirements which require careful application

In general terms, the PVC and nitrile phenolic formulations once used for sealing hem flanges have now been replaced by rubber-based and epoxy types. The former were not noted for their tolerance to process abuse, any upward fluctuations in temperature during painting leading to corrosive fumes (HCl) and/or a rigid cellular structure which could later retain moisture during service. To promote 'green strength' during manufacture and prevent movement between inners and outers, induction curing can be applied at the BIW manufacturing stage using appropriately placed/shaped platens. Careful choice of adhesive is necessary to ensure that differential distortion effects do not result in buckling/distortion of bi-metallic assemblies.

For high modulus adhesives delamination of the zinc coating for IZ coated steels can occur under some loading conditions. Structural adhesive joints may therefore be preferable with hot-dip zinc or electrogalvanized zinc coated steels.

Adhesive bonding can be considered to be particularly suitable to joining aluminium. Some adhesives can bond through mill oils and lubricants present on the aluminium component from the press shop, thus requiring no surface preparation. This is clearly the preferred choice for production planning engineers since it removes the need for surface preparation. However, by using surface preparation the choice of adhesive is expanded and durability improved (see Chapter 4). This can be seen in Fig. 6.11, where the benefits from the AVT pretreatment technology (see Chapter 4) on long-term joint strength are indicated.

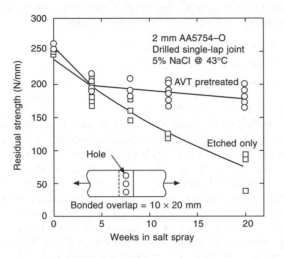

Fig. 6.11 Effect of pretreatment on long-term joint durability

Application of adhesives in the automotive industry can take place manually, but there is increasing use of automation, with dispensers ranging from the handheld variety to robotics systems. This versatility makes the use of adhesives applicable to aftermarket repair as well. One-part adhesives are the easiest to dispense requiring only a method of controlling adhesive flow and pumping equipment. Two-part adhesives which must be physically mixed to cure require additional equipment including pumping, flowmeter, mixer, etc.

The use of composite materials in semi-structural applications within a hybrid material structure is a potential technology for the future (Chapter 2). The use of adhesive bonding as a materials joining technology will be required for incorporating such parts into the structure and closures. Of the thermoplastic polymers considered for these applications, polypropylene is one of the most attractive due to its low cost, low density and good recyclability. Its major disadvantage, however, is the low surface energy which means wetting of the surface is problematic and adhesive bonding is difficult to perform without substantial assistance from surface treatment technologies including plasma, flame and chemical treatments. The mechanisms whereby a pretreatment can lead to an increase in the joint strength are:

- The creation of a surface topography for mechanical interlocking to occur
- The removal of boundary layer (or preventing its formation)
- The introduction of chemically reactive functional groups on the surface to aid compatibility with the adhesive

Investigations and research in these areas are likely to continue in the future.

6.3.1 Weldbonding

A number of deficiencies have been identified for adhesive bonded structures including limited peel strength, impact performance and temperature resistance. In contrast high peel and impact performance and insensitivity to temperature are key attributes of resistance spot welding. There is currently much attention being focused on the combination of these two processes in weldbonding.

The principle of making a good weld between two metallic sheets (either steel or aluminium), separated by adhesive, is based on squeezing the adhesive from the sheet interfaces. Work has been carried out[5] to determine the welding conditions for optimum quality weldbonded joints. The results observed for mild steel indicate that for a given welding current, the weld size obtained when weldbonding was greater than that observed when spot welding. This can be attributed to the initial higher contact resistance due to the presence of the adhesive. The major factor affecting weldability is the viscosity of the adhesive. Too high a viscosity and the adhesive cannot be displaced from the joint to allow the passage of current necessary for the formation of the weld. However, it is important to note that too low a viscosity will result in excessive movement of adhesive prior to curing and the potential for poor weldbond quality. The mechanical performance of weldbonded joints is discussed in Chapter 4.

6.4 Mechanical fastening

As an alternative to spot welding, systems have been developed using self-piercing rivets/mechanical interlocking of panels to achieve similar properties. The mechanical performance of the joints can vary with material performance, the manufacture of the joining system and joint design.

The main drawbacks associated with these processes are the limited peel strength particularly for clinched joints and the high degree of access and weight of the systems used to produce these joints. These systems need to be particularly rigid to withstand the high application forces during the fastening process and can approach five times the force level for spot welding. However, although resistance spot welding remains the major process for autobody materials assembly, mechanical fastening techniques such as spot clinching and self-pierce riveting offer benefits under certain circumstances. These include:

- The joining of material combinations which cannot be easily welded such as pre-painted steels or very dissimilar metals. For example, many of the aluminium applications in autobody structures make use of mechanical fastening techniques to overcome the poor inherent weldability of aluminium
- The joining of materials in applications where a high fatigue life is critical compared to static strength
- The joining of materials where a long tool life is of high relative importance

Mechanical assembly methods like screws and bolts have been considered as an autobody materials joining technology for some time and some excellent texts are available providing background science to these technologies.[7] In the last few decades some new systems have appeared on the market which are particularly applicable to the joining of mixed material design solutions. They can provide an efficient design solution (particularly when they are used in association with adhesive bonding) to a wide range of assembly problems. It should be appreciated that an exhaustive description of each and every mechanical fastening system is not possible in a book of this type.

Traditional riveting operations must be preceded by a piercing step, i.e. a punched hole is required before introducing the rivet. It is important to remove burrs and metallic shavings introduced by the piercing tool since these can easily lead to corrosion problems in the assembly and prevent close contact of the sheets to be joined. A flat and clean surface is necessary for optimum shear performance. Self-piercing rivets do not require such a step, since the rivet itself makes its own hole, Fig. 6.12.

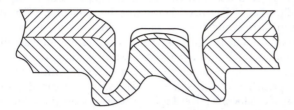

Fig. 6.12 Self-piercing riveting section

The rivets are pushed into the sheets by a small hydraulic power device with special overhang pliers. The force needed to squeeze the rivet into the sheets being high, the overhang distance is limited because of the rigidity of the pliers. Commercial systems are generally found on the market in the form of manual, easy to handle devices although there is the flexibility to incorporate the devices into automated robotic lines. Optimum performance is achieved when the joining direction is from the hard/brittle material to the soft more easily deformable material. In addition a joint should be made from the thinner to the thicker material. The aesthetic appearance of self-piercing rivets is excellent with the flat head of the rivet providing visual continuity on the sheet surface. Coloured rivets are even possible for prepainted applications.

A further alternative to spot welding is the spot clinching system. As with the self-pierce riveting system, the force needed to permanently deform the sheets is obtained through a hydraulic punch, Fig. 6.13.

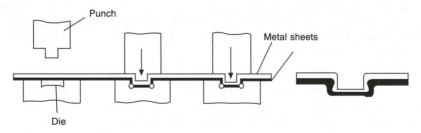

Fig. 6.13 Spot clinching process

6.5 Learning points from Chapter 6

1. Resistance welding is the predominant mode of joining used in the assembly of steel automotive body assemblies. Considerable experience exists in equipment design and manufacture and the versatility of the process lends itself to the automated and robotic application demanded to meet today's productivity targets.

2. Spot and fusion welding equipment weld settings can be relatively easily changed

to accommodate increasing body contents of high strength steel grades. For running changes within the lifespan of one model allowance must be made for the derivation of new weld lobes and provision of resetting time during non-productive periods.

3. Additional zinc coated steel body contents and conversion from one type to another (e.g. IZ to hot-dip galvanized) will require not only minor adjustment of current and force levels, but also an increased tip dressing frequency to compensate for decreased current density as the tip area is enlarged due to Cu–Zn alloying. This additional time must be allowed for in production cycles to maintain required output rates.

4. Fusion welding is still essential for many body seams and seam tracking systems are recommended for automated systems if consistent weld location is to be achieved. Fume extraction is essential for the fusion welding of coated steels.

5. Weld inspection procedures still rely mainly on the chisel test and the confirmation of the formation of a fused 'slug' meeting required dimensions. Some high strength steels (including rephosphorized and C–Mn grades) are prone to partial slug formation but allowance should be made to include the adjacent 'bright zone' of obvious fusion in the measurement of weld diameter.

6. Significant changes are required for the resistance welding of aluminium compared with steel. Higher current levels are required which necessitate larger welding transformers and spot welding electrode life is reduced to 250–300 spot intervals (before tip dressing is essential).

7. Due to limited access to internal surfaces of hydroformed sections fusion welding is probably the most process tolerant method of joining although single side spot welding appears a viable option for the future.

8. For optimum performance and durability of adhesive bonding, use of a pretreatment is recommended and robotic application used to ensure consistent location and minimize operator contact.

9. Mechanical fastening systems are finding increasing use as mixed material joints become more evident and self-piercing rivets provide one method of joining prepainted steel or aluminium.

10. Laser welding of both steel and aluminium BIW structures provides an effective method of increasing the stiffness of linear joints. Due to the need for precise beam tracking, extremely accurate panel fit-up is mandatory and the required touching contact is obtained by pressurized rotating wheels running immediately ahead of the beam.

11. Allowance should be made in design for a gap with laser joints in zinc coated steels to allow for the escape of zinc fumes which can cause expulsion of weld metal and porosity.

12. Higher rated laser power units are required for both CO_2 and YAG welding of aluminium due to reflectance and high thermal conductivity.

References

1. Williams, N.T., Private communication to R.J. Holliday.
2. Holliday R.J., 'Mechanisms of Electrode Degradation when Spot Welding Coated Steels', *Science and Technology of Welding and Joining*, Vol. 3, No. 2, 1998.
3. Natale. T.V., 'A Comparison of the Resistance Spot Weldability of Hot-dip and Electrogalvanised Sheet Steels', SAE Paper 860435.

4. Mansour, T.M., 'Ultrasonic Inspection of Spot Welds in Thin Gage Steel', *The American Society for Non-destructive Testing*, 1988.
5. Jones, T.B., Private communication to R.J. Holliday.
6. Westgate, S., 'Clinching and Self-piercing Riveting for Sheet Joining', *TWI Bulletin*, May/June 1996.
7. Messler, R.W. *Joining of Advanced Materials*, Butterworth-Heinemann, 1993.

7 Corrosion and protection of the automotive structure

Objective: To understand the main causes of corrosion affecting automotive body panels and methods of prevention and control that can be applied through design, material specification and manufacturing procedures.

Content: Types of corrosion relevant to automotive body panels – design for prevention – systems adopted for the protection of steel substrates – zinc coatings, types and methods of manufacture – paint pretreatments, electropriming, intermediate and top-coat systems – supplementary protection, wax injection and adhesive application – empirical vehicle and laboratory test methods – predictive scanning tests for early detection of corrosive activity.

7.1 Introduction

As most of the experience in volume car manufacturing and service is related to steel structures the content of this chapter tends to reflect the knowledge gained with this material. Experience with aluminium is essentially similar but where this differs with regard to corrosion mechanisms and processing this has been pointed out, where relevant, and the same applies for other materials.

Comparison of the recent performance of the body shell in resisting body corrosion shows a vastly improved performance compared with the behaviour expected even ten years ago. Although, as described in the introduction, consumer pressures and codes of practice, such as those introduced by Canada and the Nordic countries, prompted a serious re-evaluation of strategies adopted by individual companies (specially those stung by accompanying financial penalties) the main incentives have been mainly competitive. Companies are very aware of competitor practices regarding design, processes and materials utilization and particularly when these are linked with increasingly elongated durability warranties. For instance a major influence during the 1980s was the use of 100 per cent zinc coated steel for the Audi 100 models which was coupled with a warranty term of 6 years freedom from perforation and 3 years from cosmetic deterioration. Despite a 20 per cent cost premium for these materials compared with uncoated mild steel and additional manufacturing complications other companies adopted zinc coated steels and although this was often far more selective – mainly on more vulnerable panels the norm is now generally at least 70 per cent of the bodyweight in zinc coated steel. Intensive 'strip-down' exercises will have revealed better design configurations used by competitors, e.g. allowing better electoprimer ingress, differences in performance with newer coatings, or advances in supplementary materials application, e.g. cavity and underbody waxes. These issues will be covered in detail later in this chapter which will first cover the principal modes of corrosion relevant to body structures, and then successively review recommended design features and iterative vehicle test procedures, coated steel manufacture and use, and supplementary modes of protection.

7.2 Relevant corrosion processes

Although varied in terms of nature and geometry, the types of corrosion experienced in automotive situations usually have electrochemical origins. A piece of uncoated low carbon mild steel when immersed in a suitable electrolyte will create a number of local anodes and cathodes simply due to differences in composition or other physical or chemical inhomogeneities. Thus cells are formed in which an anodic oxidation reaction is proceeding according to the following equation:

$$Fe = Fe^{++} + 2e^-$$

Simultaneously a cathodic reduction process is proceeding whereby electrons are being consumed with the formation of hydroxyl ions:

$$O_2 + 2H_2O + 4e^- = 4OH^-$$

These are the anodic and cathodic partial reactions and the following overall reactions then proceed assuming a neutral, aerated, aqueous electrolyte:

For iron: $2Fe + O_2 + 2H_2O \rightarrow 2Fe(OH)_2$

later $2Fe(OH)_2 + 1/2O_2 \rightarrow 2FeO(OH)(rust) + H_2O$

For aluminium: $2Al + 3/2O_2 + 3H_2O \rightarrow 2Al(OH)_3$

Electrochemical activity will proceed with electron flow in the metal from anode to cathode (and current flow in the opposite direction) with the result that anodic dissolution at a chemical or physical discontinuity can eventually lead to perforation. Other types of cell can also exist in automotive situations, for instance in the case of bi-metallic corrosion when two metals are in contact in the presence of an electrolyte, the metal with the lower potential will become the anode with the production of electrons and associated metallic ions, the other acting as the cathode. An indication of the likely behaviour in bi-metallic situations is given by reference to a table based on relative standard electrode potentials[1] or more relevant in the current context galvanic potential measurements ('galvanic series') taken in a relevant electrolyte such as sea water. Materials of interest here are ranked in order of 'most noble' to 'least noble' and are summarized below:

Table 7.1 Galvanic Series for relevant metallic materials. (After Fontana and Greene[39])

Noble or cathodic	Platinum
	Silver
	18-8 stainless steel
	Nickel
↓	Monel (70% Ni, 30% Cu)
	Copper
	Brass
	Steel
	Aluminium
	Zinc
Active or anodic	Magnesium

Such tables can only be indicative as the degree of corrosion attack depends on other factors, as will become evident later. The rate of corrosion is affected by other influences, which can be evaluated by polarization techniques (measurements of current and potential at anode and cathodes) and depicted in the form of 'Evans diagrams'.

The products possible from corrosive reactions can be indicated from Pourbaix diagrams which show thermodynamically stable phases as a function of electrode potential and pH. Both subjects are considered in more detail later in this chapter.

Bi-metallic corrosion risks must be recognized during design and although it is tempting to mix steel and aluminium, for instance, to combine strength with low weight, precautions such as sealing and interpanel insulation with fibre washers must be taken to prevent perforation. However, another beneficial bi-metallic effect may be utilized to minimize this effect whereby galvanizing the steel (coating with zinc) may reduce the difference in potential between aluminium and steel, and the zinc sacrifices itself preferentially, while taking advantage of the slower inherent corrosion rate of the zinc. Fastener systems must be especially carefully chosen because any tendency to bi-metallic dissolution and coupling a small anode to a large cathode can concentrate anodic attack to a small area with disastrous results.

A further common cause of electrochemical corrosion within the vehicle structure is differential aeration cells such as can be created in crevices and between flat surfaces where electrolytes such as de-icing salt solution can accumulate over long periods as shown in Fig. 7.1(a).

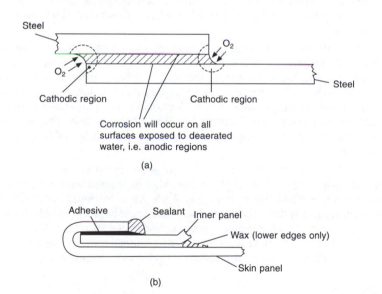

Fig. 7.1   (a) Principle of crevice corrosion or differential aeration cell.
(b) Typical seam used for door hem — traditional source of crevice attack

It is found that within such crevices a cell is formed between the cathode, which forms where oxygen is accessible, and the anode which develops at the less exposed location. Again it is the anode which eventually perforates, and Fig. 7.1(b) illustrates a section of a clinched joint showing how protection is applied in the form of a cosmetic bead and wax film applied to a hemmed seam, with the objective of excluding any harmful electrolyte. Zinc coated steel would now normally be specified for all surfaces especially within a 'wet' zone at lower regions of the bodywork.

7.2.1 Features associated with the corrosion of aluminium and other non-ferrous body materials

7.2.1.1 Aluminium alloys

Because of the basic differences in chemistry between aluminium and its alloys and steel, aluminium exhibits specific features which tend to characterize its general behaviour. First, it develops a tenacious oxide film which promotes long-term corrosion resistance and this can be grown further to form thicker protective and decorative films. This can have disadvantages, however, as this has high electrical resistance which results in a shortened electrode life when spot welding and it can require specalized mechanical or chemical pretreatments to achieve the surface states necessary for paint adhesion and retention of adhesive bond strength. Like most materials showing a passive film, aluminium sheet surfaces are prone to pitting corrosion, where the depth of attack exceeds the diameter and the bottom of the pit is anodic to the surrounding area. This can be influenced by the chemistry of the sheet or extrusion, grain boundaries, pick-up of steel or other particles from tools or on reworking, and the presence and form of intermetallic compounds. The latter are important in determining the overall corrosion resistance of particular alloys, the electrochemical potential of the intermetallic phases, and how noble they are relative to the matrix, being a critical factor. Thus copper bearing alloys can be prone to corrosion in automotive situations whereas Al–Mg and Al–Mg–Si have shown a good performance under most conditions.

Aluminium alloys of the 6xxx series can also be prone to filiform corrosion, thread-like tracks which start locally at a chip or cut-edge, propagating under the paint film, the tip of the thread behaving as an anode, the 'tail' behaving cathodically. These grow more profuse as corrosion proceeds until they cause delamination over a large area. The driving force for propagation is the electrochemical potential difference between the tip and tail caused by differences in pH and composition. The presence of chloride ions and dry sanding of the surface increase the incidence of this type of cosmetic corrosion.

Some aluminium alloys are also susceptible to stress corrosion cracking which is associated with the appearance of the β phase Mg_5Al_8 within some high magnesium alloys such as the AA 5182Al–Mg system (ca. 4.5% Mg). In moist conditions at moderate temperatures precipitation of the β phase can occur at grain boundaries and being anodic to the matrix cracking along grain boundaries can occur. Alternative alloys such as AA 5754 (3% Mg) have therefore been developed for use in applications where unsuitable conditions could be a problem – such as the engine compartment.

7.2.1.2 Magnesium

The nearest application of magnesium to the body structure prior to the cross-beam described in Chapter 2 was the gearbox cover and this normally alternated with aluminium depending on prevailing costs. Early applications developed a reputation for poor corrosion resistance but as described in Chapter 5, the higher purity variants of these alloys with lower levels of heavy metal impurities such as iron, copper and nickel have significantly improved corrosion performance. The AZ91C (Chapter 2) has now been largely replaced by higher purity variant AZ91E with a proven corrosion rate 100 times better in salt-fog tests.

Due to limited experience with magnesium alloys in autobody applications reference is made principally to aerospace technology and it is generally acknowledged that this vast improvement in corrosion resistance made recently has led to a wider application on new programmes such as the McDonnell Douglas MD500 helicopter for parts such as gearbox casings.[38] It must be noted that magnesium is positioned at one extreme (anodic) of the galvanic series and for this reason bi-metallic situations such as the use of mechanical fastenings must be assessed very carefully.

7.2.1.3 Polymers

As for joining of plastics the number of different polymer combinations makes any generalization regarding corrosion difficult and the reader is directed to specialist publications such as *Engineering Materials 2*,[41] or TWI for further guidance. A very useful chart on 'Plastics vs Environmental Factors' has been compiled by Fontana and Greene.[39]

7.2.3 Mechanism of paint degradation

Reference has so far been made to corrosive mechanisms relative to metallic surfaces and joints, but of equal importance is the performance of the protective paint film. Harmful effects leading to deterioration of the film integrity must be understood and remedial measures taken wherever possible. The major cause of adhesion loss as shown by Dickie and co-workers at the Ford Motor Company,[2] and Leidheiser and Wang,[3] is due to the breakdown of the organic layers by hydroxyl ions produced by the cathodic reaction. This process of hydrolysis or saponification can be expressed[4] by the following equation in a coating containing ester groups:

$$R_1{-}COOR_2 + OH^- \quad > \quad R_1COO^- + R_2OH$$
ester $\quad$ + hydroxyl $\quad$ carboxylate + $\quad$ alcohol

The progressive breakdown of the paint film at the point of a defect on plain mild steel and a zinc coated steel surface, showing chemical and physical causal factors, is illustrated in Fig. 7.2.

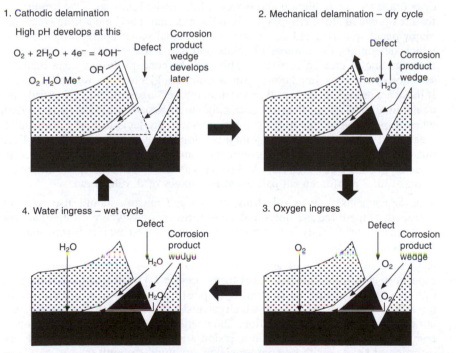

Fig. 7.2 Schematic breakdown of disbondment of automotive paint film subject to cyclic dry–wet conditions during service. (After Granata[5])

Separation of the reactants is a key factor but obviously prevention of corrosion is paramount and as well as the supplementary methods which can be used to achieve this by coatings and sealants, to be covered later, fundamental consideration must be given to basic styling and design features. These can have a significant effect on the prevention of the occurrence of trouble spots such as mud traps, other elemental ingress and crevices. This can only come from historical experience accrued from expert analysis of service inspections, detailed appraisal of competitor designs and more specific iterative tests as individual model development proceeds. These are of necessity accelerated procedures but devised to incorporate extremes of climatic and service conditions and are normally based on 12 week or longer vehicle test specifications which comprise successive gravel chipping, salt spray and mud accumulation stages interspersed with daily road running and nightly humidity chamber exposure as described later.

As stated earlier, sympathetic design is a key factor and various principles have been applied vigorously over recent years to ensure that optimum durability is obtained from panels produced in increasingly light/thin materials and these are summarized as follows.

7.3 Effective design principles

These can be divided into influences arising from styling features, subassembly, panel and design and associated production process practices.

7.3.1 Styling

Apart from aesthetic and aerodynamic influences the general body shape must discourage the accumulation and ingress of debris and minimize the prominence of features prone to stone chipping. It is at the external 'clay' modelling stage that the major panel split is decided, i.e. where the actual panel configuration is planned which determines the number of joints and associated seams, all of which are a potential site of crevice corrosion. This and the concept design stage includes the selection of materials for adjoining panels which could introduce bi-metallic interfaces. If these are deemed necessary then location in 'wet' areas should be avoided. As well as aluminium and magnesium this also applies to zinc coated/mild steel mixed interfaces where saponification due to the cathodic reaction products from the bi-metallic cell created (at galvanized/uncoated joints) is a long-term risk. Panel splits or divisions must be minimized and parts consolidation must be encouraged, e.g. larger mono-side designs, one-piece doors, etc. Add-on parts should be avoided despite the temptation to make local reinforcement patches at late stages of development.

Trim attachments such as sidestrips, badges and mirrors should also be carefully designed, as these require holes and again form crevices where differential aeration cells can initiate. Adhesive attachment is to be encouraged but if holes are unavoidable these must be inserted prior to paint.

Regarding development of assemblies inboard of the external panels, mud traps especially in locations behind wheel-throw areas must be avoided and if they cannot be reconfigured should be covered by a plastic moulding, e.g. wheel arch liners, underfloor deflectors/covers, etc. which can resist abrasion by stone chipping, obscure cut edges and prevent accumulation. The overall design must also permit ingress of paint during electrocoat primer immersion and box section waxes, usually by the insertion of holes – without compromising structural strength.

Associated processing should ensure a minimum of fusion welding with attendant burn-off of zinc and/or deposition of iron oxide scale. Bi-metallic assembly designs, e.g. aluminium fenders on a steel substructure, should ensure adequate insulation by use of sealers (strip or manually applied) or suitable gasket material. Sealers should be consistently applied – preferably by robot where exact placement can be programmed and/or modified – and not be subject to process abuse, e.g. overheating during curing which could cause cracking and subsequent retention of moisture at the substrate. This can happen with some nitrile-phenolic and PVC formulations where overstoving can produce a cellular structure which will retain moisture later in service, and acidic fumes can be released on breakdown of PVC.

7.3.2 Subassemblies

One of the key problems of subassembly design is avoiding complexity because as well as complicating the joining processes, panel coverage by paint, waxes and wax injectants is inevitably impaired. If reinforcements are necessary on internal sections it is therefore a basic requirement that circulation is maximized by redesign and iterative testing to confirm efficiency of surface treatments (see later). The real dangers are touching surfaces, which as well as causing paint depletion promote crevice corrosion.

Steps taken to improve paint coverage include the incorporation of fluted flanges (castellated sections 1–1.5 mm deep), which allow easier ingress of primer. Waxes have been developed to maximize spread without excessive drippage in production areas where these are applied. Ideally the hot wax application should be adopted as introduced by Audi, whereby injection is carried out under pressure into all underbody box sections thus forcing wax into most orifices and seams and retaining a thick film when the formulation cools.

7.3.3 Panels

Individual panel design features should include drain holes located at the lowest extremity of the section in subsequent parent subassemblies. Process related features must include protection by oils approved to give at least 3 months' protection in the press shop, storage and assembly. Panels must be burr free to limit damage in transportation, prevent localized contact with bi-metallic assemblies, and ensure paint depletion does not occur at panel edges, giving premature rust bleeding at cut-edges. It is emphasized that in the case of bolt-on panels such as aluminium front wings, attachment directly to steel substructures, if unavoidable, should make use of galvanized nuts, bolts and washers, or efficient use of gasket materials to prevent longer-term perforation resulting from bi-metallic corrosion. Aluminium skins attached to steel inner structures should be separated by a uniformly applied layer of adhesive or appropriate sealer.

Inevitably, due to compromise with other design criteria, e.g. reinforcements within sections of front longitudinal panels to meet impact resistance – which may conflict with the size/number of electroprimer access holes or create contacting surfaces, paint coverage may be impaired. To counter such weaknesses therefore it is necessary to use supplementary protective systems comprising sprayed organic or polymer compounds. These are briefly described in the ensuing section following a more detailed description of the primary metallic, paint and pretreatment protection systems.

7.4 Materials used for protection of the body structure

7.4.1 Zinc coated steels, types and use for automotive construction

As briefly described previously the use of zinc to protect steel depends on its preferential galvanic dissolution giving cathodic protection of any exposed steel substrate, in the presence of a suitable electrolyte. This is further reinforced by the barrier effect provided by the zinc coating itself, which corrodes 10–100 times more slowly than steel, and deposition of the insoluble corrosion products $Zn(OH)_2$ and $ZnCO_3$.[6]

7.4.1.1 Mode of application

The mode of application chosen is dependent on cost, quality of finish, and also ease of processing for mass production purposes, but it is generally accepted that fabricating steel sheet pregalvanized at the steel plant by hot dipping or electrogalvanizing of the continuous strip, is the most convenient method. The post assembly zinc dipping process such as that used for the Renault Espace subframe offers the advantage of increased torsional stiffness but the slow throughput, variable surface finish, need to control heat distortion, and added weight preclude this type of treatment from normal high volume production (e.g. 4000 units per week). Although theoretically possible, electrogalvanizing of larger automotive assemblies is currently considered impractical due to the complexity and logistical problems surrounding the various processing stages and difficulty of depositing ('throwing') a film of uniform thickness into box sections and crevices featured by a typical body. Energy costs would also be higher.

The emergence of zinc precoated steel sheet as the significant body panel material is comparatively recent (1980s onwards),[7] although it has been used selectively to confer galvanic protection on luxury cars such as the Rolls-Royce since the 1960s. The Silver Spirit has featured relatively thick coatings of 275 g/sq. m for most underframe parts but this has been used to achieve the exceptional level of durability associated with this prestigious vehicle and since this model is made in low volumes, production time can be allowed to manually dress copper welding electrodes (needed to cope with the formation of brass during spot welding of heavily zinc coated sheet) and optimize surface finish. Additional material costs of £50 per body could also be more easily accommodated within a showroom cost of £90 000 rather than the £20 000 (or lower) purchase price of a higher volume model. Initially automotive zinc coatings were relatively unsophisticated, lacking control of thickness and surface finish.

The consumer and competitor pressures responsible for the upsurge in the use of galvanized steel, which was experienced in the 1980s, have already been referred to. A further factor, which contributed to this, however, was the coincident uplift in quality of the coated products. A much wider range of galvanized sheet products now became available from technologically advanced European continuous strip production lines (both electrogalvanized and hot dip) offering coating specifications which recognized the special requirements of the automotive producer (surface finish, formability, thickness control, etc.). A brief summary of the main coating types is presented in Table 7.2 while a more detailed analysis of pros and cons is given in Table 7.3.

These products were initially developed by the American steel companies who responded earlier than Europe[8] to the need for improved protection for vehicles being driven in the winter environment around the Great Lakes. Also the Japanese who were beginning to realize the commercial opportunities that exist for developing more sophisticated,

Table 7.2 Zinc coating types used for manufacture of automotive body panels

Type/Typical thickness (μm)	Advantages	Disadvantages
Electrogalvanized ZE 7.5	Good surface, formability	High cost (hot-dip galvanized + 10%)
Hot-dip galvanized Z 7.0–20.0	Good drawability, cost	Electrode wear
Hot-dip galvanized ZF 6.0 Fe–Zn alloy coated	Formability, weldability, cost	Slight powdering in compression
Duplex Zn–Ni + organic primer 3.0 1.0	Formability, paintability, corrosion	High cost, restricted availability
Duplex Zn + organic primer 5.0 3.0	Reduced wax in box sections and reliance on sealer application in hem seams	High cost, increased electrode tip dressing frequency

value-added types of coating, the technology for which could be passed on to European producers, partly to feed the Japanese satellite car plants now emerging (e.g. Honda, Toyota and Nissan in the UK). Types of steel available included single and differential coating thickness options, alloy formulations which helped overcome welding difficulties, and matte surfaces to improve paint finish.

Apart from Audi, other competitors realized that the improved performance being achieved with zinc coatings was being recognized by the consumer and highlighted by annual publications and moved to ensure that at least equivalent levels of durability were being achieved. As described more recently[12] most prominent car producers adopted coated steels, although the proportion and type of coating differed. Among European-based manufacturers preferences can be summarized as shown in Table 7.3.

Table 7.3 Use of different coating types by major car manufacture

Coating type	Manufacturer
Hot-dip galvanized (Z):	Volvo (inner panels), PSA, Audi (inner)
Galvanneal (ZF):	Rover, Honda, Toyota, Ford
Electrogalvanized (ZE):	BMW, Fiat, Renault, Audi and Jaguar
Duplex/alloy ZE:	Nissan (Durasteel), GM Opel (Zn–Ni)

The planning assumption made by these carmakers was that the market now required 5 years' freedom from cosmetic corrosion and 10 years from perforation. Coatings of 7.5 microns are normally used to achieve this competitive performance although with indications in 2000 that 12 years' freedom from perforation will become the norm very shortly (as offered on the GM Astra and VW Golf from 1998), some panels may require 10 microns.

7.4.1.2 Development of a car manufacturer's coated steel policy

With the background of extended warranties, up to 90 per cent of the body panel content of a typical European vehicle is now specified in zinc coated steel as illustrated in Fig. 7.3.

Mild steel, uncoated, is maintained for upper panels as this upper zone is regarded as a primarily 'dry' area, remote from the more critical underbody corrosion prone areas. Ideally, the main objective is to use one type of coating for all applications in

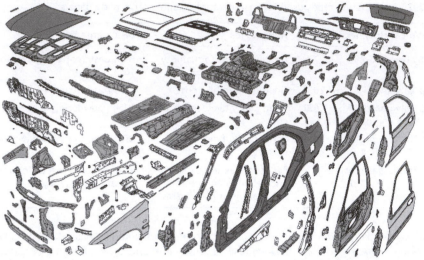

Ez (electrogalvanized)　Durasteel　IZ (galvanneal)

Fig. 7.3 Application of coated steel within a typical body structure

potentially wet areas to simplify logistics, reduce costs and improve processing consistency. This should satisfy both internal and external standards and provide optimum weldability, paintability and corrosion resistance subject to the preferences and facilities of individual manufacturers. To help focus on these issues and examine opportunities for rationalization it has been found useful to compare the performance of different product types according to key user criteria, as shown in Table 7.4.

As product development takes place so the ratings in the table change and while galvanneal, the iron–zinc alloy coating ('ZF' designation according to Euronorm 10142), was favoured by a number of major car producers at the time the table was issued (1998) decreasing cost and increasing quality now once more make the free zinc hot-dip ('Z') coating an attractive option – and an alternative to electrogalvanized sheet and Durasteel (duplex organic primer coated Zn–Ni) for many manufacturers.

7.4.1.3 Hot-dip galvanizing process

Some historical detail may be useful in understanding progress made with these coatings and in the 1930s the primary supply source for hot-dip products were Sendzimir lines or similar as shown schematically in Fig. 7.4.

These allowed the economies of coil production to be exploited but the properties of the in-line annealed strip were moderate and variable. Developments such as post annealing gave some improvement in ductility over the standard product which suffered from limited annealing time at temperature, but the product was still not suited to the increasingly demanding quality standards of the vehicle manufacturer. The emergence of new lines such as the Corus Zodiac galvanizing line shown schematically in Fig. 7.5 really paved the way for the widespread specification of zinc coated steel panels, as users were now assured of

- Substrates which could utilize IF forming and high strength technology to give much improved ductility levels
- Coatings with consistent thickness levels imparted through nitrogen gas knife technology and increased bath surveillance, allied to improvements in adhesion arising from Nb–Ti steel additions

Table 7.4 Chart used to compare the overall characteristics of different zinc coating types[8]

Issue 2 June 2000

Ranking factor	Body skin panels					Body interior panels				
	Weighting	EZ	IZ	HDG	Zn/Ni	Weighting	EZ	IZ	HDG	Zn/Ni
Corrosion resistance	10	5	5	5	2	10	5	5	5	2
Cost	10	3	4	5	1	10	3	4	5	1
Health & safety (weld/rework)	10	5	5	5	1	10	5	5	5	1
Availability (No. of producers)	8	5	5	5	3	8	5	5	5	3
Surface finish (quality control)	8	5	4	4	5	3	5	4	4	5
Formability (powdering/adhesion)	8	4	3	5	2	8	4	3	5	2
Coating thickness control	6	5	4	4	5	6	5	4	4	5
Press tool wear	5	5	4	4	2	5	5	4	4	2
Weld electrode life	5	3	5	3	5	8	3	5	3	5
Number of users	4	5	5	3	2	4	3	5	3	2
Effect on substrate formability	3	4	4	4	4	3	4	4	4	4
Paintability (e-coat cratering)	3	5	4	5	5	3	5	4	5	5
Effect on fusion welding	1	3	4	3	4	3	3	4	3	4
Coating adhesion (with adhesives)	1	5	3	5	1	5	5	3	5	1
Weighted score		367	356	368	225		369	371	383	227

EZ = Electrozinc EN 10152 (+ZE)
IZ = Galvannealed EN 10142 (+ZF)
HDG = Hot-dip galvanized EN 10142 (+Z)
Zn/Ni = Zinc nickel
Scoring:
5 = Excellent
4 = Good
3 = Average
2 = Acceptable
1 = Poor
* Subject to E-coat plant modification

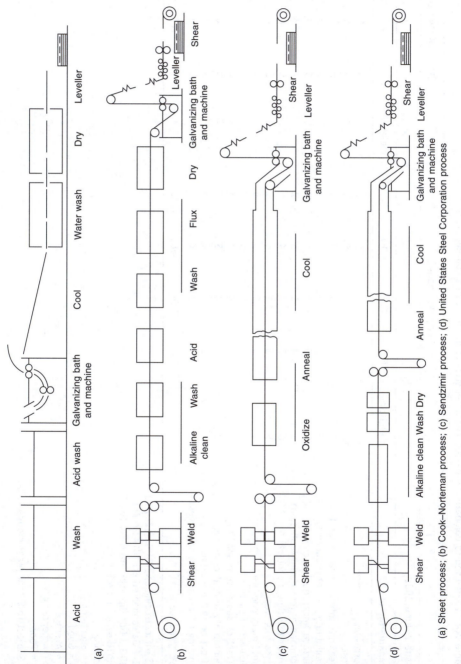

(a) Sheet process; (b) Cook–Norteman process; (c) Sendzimir process; (d) United States Steel Corporation process

Fig. 7.4 Schematic layout of continuous galvanizing lines 1930–1960. (After Edwards[9])

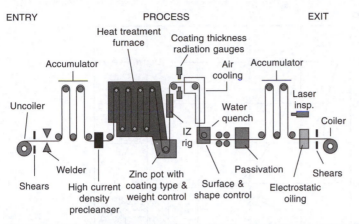

ENTRY   PROCESS   EXIT

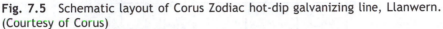

Fig. 7.5 Schematic layout of Corus Zodiac hot-dip galvanizing line, Llanwern. (Courtesy of Corus)

These more modern processes are termed continuous annealing galvanizing lines (CGL) and have the capability to fully – rather than partially – anneal the substrate, and with interstitial-free technology (vacuum degassing plus titanium (niobium additions) can match the properties achieved with uncoated steels. However, it is important to differentiate the processing of these forming grades from high strength variants and recognize that although cooling rates are rapid they cannot match those achievable with CAPL lines used to continuously anneal uncoated strip. Here cooling rates of 30–60°C/s can be attained and the martensitic phase obtained if required, e.g. with dual-phase steels. With CGLs, cooling rates are slightly lower which means that only bainitic or partially bainitic structures may be obtained unless the composition is modified, e.g. by Cr or Mn additions, and this in turn may require modified joining or finishing conditions. This highlights another complication arising from the use of more esoteric steels, namely the differences in performance (e.g. in welding response) that can arise from nominally the same product from different suppliers, due to varying product specifications and processing, and confirmatory trials may then be necessary for such steels, e.g. dual phase.

To clarify the difference between the Z and ZF coatings, both are produced on a continuous hot-dip galvanizing line as shown schematically in Fig. 7.5, the alloy product undergoing a post dipping heat treatment under carefully controlled cooling conditions to promote favourable phase formation.

7.4.1.4 Relative performance of zinc coatings

The alloy coating is still popular where emphasis is on manufacturing (good weldability, paintability) especially with the Japanese UK car plants. Progressive improvements in the coating adhesion and surface finish now make this coating a proposition for outer as well as inner panels. Initially, this seemed a surprising choice for press formed panels due to the brittle nature of the alloy system comprising zeta, gamma, and delta phases, as shown in the phase diagram Fig. 7.6, developing hardness values up to HV420. However, experience quickly confirmed reasonable ductility and coating adherence plus corrosion resistance which matched that of electrogalvanized steel in underbody assessment tests over 100 000 miles. The choice of galvanneal is supported by the reports of other workers. Townsend[10] presents results following comparison of most types of coating in undervehicle conditions showing iron–zinc alloy coatings, together with Zn–Ni to outperform plain zinc – while Johannson and Rendahl[11]

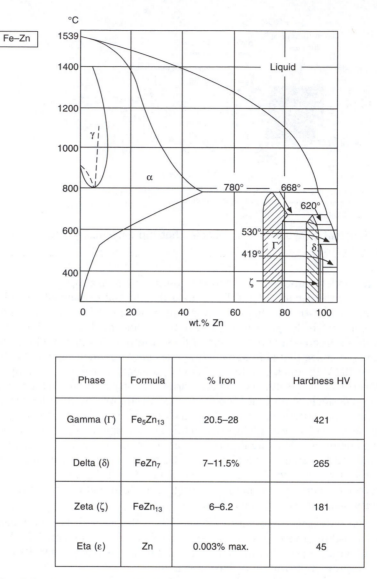

Fig. 7.6 Iron—zinc constitutional diagram and properties of phases

Phase	Formula	% Iron	Hardness HV
Gamma (Γ)	Fe_5Zn_{13}	20.5–28	421
Delta (δ)	$FeZn_7$	7–11.5%	265
Zeta (ζ)	$FeZn_{13}$	6–6.2	181
Eta (ε)	Zn	0.003% max.	45

showed similar performance in marine and mobile tests. Miyoshi[12] and co-workers at Nippon Steel also found a better performance for IZ over free zinc coatings and attribute this to the better paint adhesion experienced with the fissured/rougher surface and a higher resistance to breakdown of the iron—zinc phosphate layer by hydroxyl ions from the cathodic reaction. This is further endorsed by Lee and Hiam[13] who carried out electrode potential (SCE) measurements on each of the zeta, delta and gamma alloy layers and found these to be different (–870, –820 and –770 millivolts respectively) and less negative, in a solution of sodium chloride and zinc sulphate, than a hot-dip galvanized coating (–1.03 millivolts). The better performance was related to the more noble less reactive nature of the alloy and again better adhesive bonding. However, this conflicts with the choice of other major car manufacturers such as BMW and Peugeot who prefer free zinc coatings, as cut-edge protection is claimed to be more efficient and the white corrosion products are adjudged preferable to the misleading red colouration observed with galvanneal in accelerated tests (not as obvious under longer-term outdoor exposure in service).

As stated, process improvements to the general hot-dip galvanizing line and particularly bath discipline have elevated the position of the extra smooth quality of the basic product to rank alongside IZ and EZ. Although not showing the same weldability as IZ in terms of settings and electrode life, equipment can generally accommodate the free zinc variant, by the use of increased currents and more frequent tip dressing, to exploit the lower costs. While consistency of surface has to be proven for outer panels, this has been demonstrated for application within the French car industry[14] and significant supplier activity is evident in improvement of zinc bath design and operational disciplines.

7.4.1.5 Electrogalvanized steel sheet and other variants

Electrozinc (EZ) had been the main European choice of coating until about 7 years ago, as the formability of normal cold rolled uncoated steel grades could be retained through the low temperature electrodeposition process, and was combined with reasonable weldability and paintability. More importantly, a consistent surface quality could be maintained in terms of both freedom from blemishes and the surface roughness/peak count requirement necessary for internal and external finishes. It is confidence in achieving the latter requirement that has led to the retention of EZ for complex exterior panels (e.g. the sidepanels shown in Fig. 7.3) in either double or single side coated condition – a further advantage of this process being the ability to deposit on one side only, by placing anodes adjacent to only one surface of the strip. A major factor causing a re-evaluation of the application of this product, however, is cost (10 per cent higher historically than hot-dipped sheet), and this can be coupled with a current undercapacity in supply. It is also claimed that organically coated zinc–nickel electrodeposited coatings (e.g. Durasteel) are less prone to saponification effects in hemmed seams and provide more effective protection in crevices and at cut-edges (blister formation due to attack of the paint by hydroxyl ions described more fully earlier in this chapter). A further advantage is that a number of cavity wax and sealer operations (and associated robotic processing/automation) may be dispensed with. Thus it has been specified for critical door outer panels (to provide internal protection to the lower seam where ingress of primer is limited) and front wings where cut-edges are not protected by clinching, i.e. bent flat as for the peripheral seams around bonnets and doors. Other zinc coating specifications adopted by European and Japanese manufacturers include electrogalvanized thinner single layer metallic alloy coatings such as zinc–nickel, where coatings have been reduced to 2–3 microns with benefits in forming and welding. Recently emerging are thicker coatings of the 'Bonazinc' family which superimpose up to 5 micron prepainted primer coatings on top of 5 microns of electrodeposited zinc. The Daimler-Chrysler 'S' Class is an example where this type of organic coating has been specified.[15] One important group of coatings under development for the future (introduction in 5–10 years) is manufactured using physical vapour deposition techniques and these are claimed to show significantly enhanced corrosion resistance[16] at lower thickness and provide a unique method of producing alloy, sandwich and gradient type coatings.

7.4.1.6 Other factors affecting performance

Relative areas of anode and cathode

The efficiency of the galvanic action provided by the zinc is related to the size of the current flow between zinc and steel but more specifically current density, and this must be related to the amount of surface exposed to the electrolyte or corrosive medium. Practical instances where this is of importance are as follows.

(i) Thickness vs durability. Ford[17] and Hoesch[18] have published data suggesting that 4 micrometres is the minimum zinc coating thickness required to maintain satisfactory in-service performance, from the limited thickness ranges tested. Other information on optimum coating thickness is surprisingly scarce although recent investigations are beginning to report back on 10 year actual automobile tests in Okinawa[19] where 40 g/sq. m IZ material just meets the 2 mm scribe scab criterion, and on-car perforation tests in Detroit (still in progress).[20]

(ii) Zinc depleted zones. Little information is available on the effect of welding and resulting areas of zinc depletion, especially for tailored laser welded blanks which feature weld zones from which the zinc has been removed. As described by Waddell and Davies[21] this is a growing technology for the production of pressed blanks with differential specifications, e.g. thick/thin, coated/uncoated, high strength/low strength combinations. Also it is necessary to confirm the corrosion resistance of areas of the body-in-white (body structure prior to paint) where the coating has been disked/filed off prior to paint. A bi-metallic boundary is formed and the longer term effects of potential activity under paint films, especially those where pretreatment or other coats have been inefficiently applied, is yet to be confirmed.

(iii) Cut-edge behaviour. The difference in cut-edge protection between galvanneal and electrogalvanized steel sheet has been studied by Suzuki et al.[22] and the latter was found to be more effective in this situation due to greater anodic activity at the corner position. Cut-edges in organic prepainted steel have been studied recently by Worsley et al.[23] using SVRET (electrochemical vibrating probe techniques, described later) and it is interesting to note that the two anodic sites at the steel–zinc interface produce symmetrical currents with paint films of equal thickness. However, when thicknesses were asymmetrical, a higher current was produced at the thinner coated interface, indicating a higher corrosion rate due to the differential aeration effect. This may be significant in automotive situations where differential situations exist, e.g. box sections depleted of paint on the inner face, and may be critical when damaged or drilled exposing the interface.

As the processing and quality aspects of the various coatings grow more similar, together with the prospect of longer anti-corrosion warranties, more organizations are refocusing on corrosion resistance for which associated test methods are critical and these will be considered next. A review of current assessment[24] procedures highlights several problems associated with available methods. Actual real-time tests are sometimes criticized as being too long for automotive design and all companies have devised accelerated procedures aimed at simulating typical extreme service conditions.

7.4.2 Painting of the automotive body structure

7.4.2.1 Introduction

Until recently the protection of uncoated steel panels was totally reliant on the pre-treatment and subsequent paint layers. However, even with zinc coatings the effect of paint is critical as the zinc merely retards corrosion and additional further bi-metallic effects can arise from any mixed zinc–iron interfaces created during assembly. Therefore it is important to understand the current status of paint processing and failure mechanisms and how this may be affected by zinc coated substrates and, increasingly, the presence of aluminium as bolt-on panels or even complete structures.

Current 'best practice' in automotive body painting can be illustrated by reference to a typical UK paint line used for the finishing of high volume medium sized cars, and this continuous process usually comprises six essential stages, namely:

- Pretreatment – featuring full dip phosphating
- Electropriming – cathodic application of a film 25–30 microns thick
- Surfacer application – an intermediate spray coat to 30–40 microns thick
- Anti-chip primer coating – applied wet-on-wet to specific areas only (20 microns)
- Base colour application – final top-coat application, 15–25 microns thick
- Clear coat finish to enhance lustre and depth of colour, 35–50 microns thick

Each of these processes will be described in more detail below but from a corrosion standpoint the initial pretreatment stage is the most critical and will be analysed in most depth.

7.4.2.2   Pretreatment

The most effective current processes are based on the following sequence of operations and extend to nine stages as shown schematically in Fig. 7.5. The most important of these is the full dip phosphate stage which ensures that full body coverage is obtained.

The painting of IZ coated steel is easier than hot-dip or electrogalvanized finishes and this is usually attributed to the presence of fissures which are developed on cooling by contraction of the coating after alloying and which act as anchoring points for the electroprimer. Hence adhesion is improved. However, it is also essential that thorough precleaning and other pretreatment procedures are carried out and the complete current process comprises nine successive stages as stated. Critical steps are the full immersion in tricationic phosphate containing nickel, manganese and zinc followed by a chrome rinse to develop the correct, fine granular phosphate structure that optimizes adhesion. The Ni and Mn additions help develop the correct proportion of phosphophyllite to hopeite and the chromate counters the effects of hydroxyl ions at defects and both inhibit the progression of the delamination front. As stated above, the optimum structure is reported to be a fine granular crystal formation rather than a coarse needle array because this is associated with a high level of porosity. Also, resistance to alkaline attack is reported to be increased due to the presence of a high proportion of phosphophyllite $Zn_2Fe(PO_4)_2.4H_2O$ compared with hopeite $Zn_3(PO_4)_2.4H_2O$, and the 'P' ratio, relating the relative presence of both, is closely monitored in relation to finished paint quality.

Pretreatments are described by Gehmaker[25] which are applicable to mixed metal structures of zinc coated steel and aluminium, these tricationic phosphating formulations containing controlled dosages of fluoride to promote uniform crystalline coatings on the aluminium surface. The quality obtained is said to be comparable to that obtained with chromating without it being necessary to pickle the surface prior to conversion. This treatment is reported to be especially beneficial for the AlMgSi alloys used for outer panels as the uniform crystalline phosphate coating slows down filiform corrosion, characteristic of painted aluminium exposed to salt, moderate temperatures and relative humidities of 60–90 per cent. Improved performance is shown for the zirconium fluoride treatment over chromate equivalents and other zirconium–titanium-based formulations are now being generally recognized as environmentally acceptable alternatives to the chromate treatment.

7.4.2.3 Electropriming

Prior to the late 1960s priming was carried out by spraying or dipping and from inspection of scrapped vehicles both processes were found to be inefficient due to lack of penetration into box sections and seams. Attempts were made to improve coverage by more open design of flanges and also by the development of improved application methods to automotive body structures such as 'Slipper-dip' and 'Rotadip' processes whereby the body was partially dipped into a bath of primer to improve protection of the understructure or actually rotated on its longitudinal axis to extend coverage to the entire body. However, inefficiencies were still evident with bodies prone to sags and runs and paint depleted areas where occluded air pockets prevented ingress of primer. Therefore there existed a need for an alternative system which ensured complete surface deposition that could meet production volumes on a continuous basis and which would improve material utilization, so electrophoresis proved a timely development. This essentially comprised the attraction of charged paint particles suspended in aqueous solution to an oppositely charged pretreated body and the subsequent crosslinking of the thermosetting paint formulation during oven stoving to provide a uniform, adherent film. To ensure maximum throwing power within all body sections it was essential to design ingress holes for the paint – without compromising strength. The location of such holes and later sealing with plugs is a critical part of the body design process and optimization proceeds until the initial production stage is reached.

In July 1961, Ford commenced operation of its first electropainting facility for wheels followed by its Wixsom plant for full bodies in 1963. Coincident development within the Pressed Steel Company at Oxford resulted in the application of the anodic process using both hydroxyl- and amine-based formulations. Due to better penetration and improved corrosion resistance (especially with bi-metallic coated steel joints), cathodic processing, whereby the workpiece is made the cathode, has now become the norm and the ASTM B117 salt spray performance improved from 240 hours to 1000 hours without deterioration. A typical cathodic electropaint bath consists of four main ingredients:

- An aqueous polymer solubilized with acid
- Pigments wet with special solubilized resin
- Coalescing solvents
- Deionized water

Typically the resin system is a complex mixture of specially adapted epoxy-polyurethane technology which is water soluble, as are the miscible glycol ether solvents which aid dispersibility and promote flowing together of the resin particles in the deposited film thereby allowing a uniform film to form. The use of deionized water is essential because of the serious effect of contaminants on electrolysis. The electroprimer film is usually 25 microns thick after curing at 180°C for 20 minutes.

Bath conditions are particularly critical for the electropriming of IZ for if the voltage applied to the body is too high on entry to the bath then cratering due to the evolution of hydrogen can occur. This happens because a discharge takes place across hydrogen bubbles and the paint particles, causing premature curing which appears as pitting of the surface ('cratering'). The voltage at which this occurs is lower for IZ than for free zinc coatings and the effect is normally overcome by adjustment of local bath conditions or use of 'high build electrocoats' – formulations which effectively stifle the evolution of hydrogen with a high solids/solvent ratio.

7.4.2.4 Surfacer

The function of the surfacer is to obliterate any adverse topographical features or local scratches/defects arising from the electrocoat stage and is normally a relatively thick polyester coat applied by spraying within the range 30–40 microns. This allows extensive rework by wet or dry sanding if required. Adhesion to electrocoat is important as this provides a contribution to overall anti-chip performance and cosmetic deterioration through substrate exposure.

7.4.2.5 Anti-chip protection

To protect forward sloping panel profiles, which are subject to damage by stone chips thrown up in the 'wheel throw area' or by preceding vehicles, it is now common practice to apply a zone of resilient polyurethane or elastomeric material, to absorb the impact energy, on top of the normal surfacer. This is approximately 20 microns thick in a non-pigmented formulation applied without the previous coat fully drying, i.e. 'wet-on-wet'. This is typically applied to zones approximately 25–50 mm wide at the bonnet front and nose of front wings.

7.4.2.6 Base colour coat

The final colour is determined by the electrostatic application of an acrylic and melamine solvent-based film 15–25 microns thick. Due to the presence of high volatile compounds and emerging emission control regulations, solvent-based formulations are now being replaced by water-based paints which are more complex latex-based solutions, requiring more expensive stainless steel corrosion resistant plant. The colour coat can be a plain pigment formulation or contain aluminium flakes to give a 'metallic' finish. Mica additions can also be made to give a pearlescent effect.

7.4.2.7 Clear coat

Due to increasing efforts to impress the customer with high image clarity, highly lustrous showroom finishes, a further acrylic clear coat is finally applied to a thickness of 35–50 microns.

7.4.3 Supplementary protective systems

As well as the more universal protection modes described above it is also necessary to reinforce the coverage given to traditionally weak areas such as the inside of box sections and vulnerable underfloor zones. Underbody seam sealing using PVC formulations is commonly carried out after final painting, using robotic application for higher volume production, to further prevent the ingress of corrosive media into seams prior to undersealing of the body. Materials have to fulfil a number of functions and on the prestige models filled bitumastic formulations have been preferred, applied in thick layers to impart anti-drumming properties. However, cheaper versions of this type applied to volume car production were prone to cracking and retention of moisture, and current models tend to favour low weight cellular PVC layers robotically applied over normal seam sealing and subsequently cured. For box sections (e.g. sills, longitudinal members, etc.), wax injection techniques of different types are used, ranging from hot waxes which are used to flood the underbody structural members internally under high pressure, to the more usual thinner formulations which are applied by lance injection methods at the end of the body paint finishing line. Ideally this latter type of wax is thin enough when introduced to the body structure to penetrate all sections and seams but without dripping from the structure

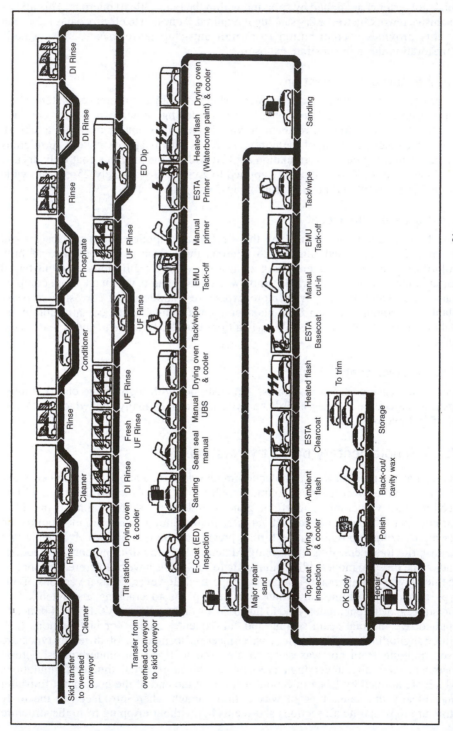

Fig. 7.7 Typical process flow chart for a 'state-of-the-art' paint shop. (After Bucholz[26])

and creating local housekeeping problems on the production line. The method of application is also critical, as if the movement of the nozzle head is not carefully planned only local deposition occurs and coverage is severely limited. Hence the revolving nozzle type application is preferred (with spiralling action on withdrawal) and this has proved more effective but is still prone to operational errors and must therefore be regularly monitored to ensure maximum efficiency. The latest injectant formulations are water based with a solids level of approximately 60 per cent which comprise waxes and corrosion inhibitors.

In the hot wax method, pioneered by the VW/Audi organization, the wax is injected hot and under pressure, so that fluidity and penetration are maximized. On cooling the material solidifies as a substantial film which provides long life to components. This method, although extremely effective, requires significant capital investment.

7.5 Empirical vehicle and laboratory comparisons

7.5.1 Vehicle assessments

To confirm and monitor an effective materials policy most automotive manufacturers now carry out two types of evaluation, the first is a running vehicle assessment involving 12 weeks' exposure to regular, alternating, salt spray, and humidity cycles. Model development schedules comprise alternate daily road running sessions and overnight storage periods in humidity ovens, which – interspersed with stone chipping, salt spray and mud packing to develop a poultice effect in vulnerable areas – last approximately 12 weeks. The timescale is often uncertain due to failure of running gear under exceptionally severe conditions, requiring replacement of mechanical parts. Hence a typical programme can exceed 20 weeks with a further strip down period required for full examination of box section interiors and seams. A schematic sequence is shown in Fig. 7.8.

Such tests are expensive and time consuming but are useful in identifying weak areas which can then be redesigned, and further iterations carried out. As hinted, some observations can prove misleading in terms of actual behaviour, e.g. IZ appears to have a considerably longer life in service, even when not well protected, than would be predicted from accelerated tests, perhaps due to the formation of a natural patina. Accelerated tests quickly show red rusting probably due to preferential attack of iron concentrations in the coating whereas in atmospheric staining rusting appears less marked possibly due to the protective film of natural oxides and carbonates.

7.5.2 Laboratory tests

These empirical vehicle corrosion tests are supplemented by the second type of evaluation, laboratory or humidity tests such as the well-known ASTM B117 specification and more recently developed cycling corrosion tests, which seek to vary extremes of environment under more controlled conditions for periods of up to 6 weeks.

Despite disadvantages, the above current methods of assessment are still universally used to compare automotive materials, and increasingly efforts are being made to more realistically correlate laboratory with environmental extremes. It is also important, however, to understand the basic mechanisms responsible for various types of corrosive attack and again it is important to be able to accelerate situations and compress investigations within the timescale of normal laboratory research programmes. In the

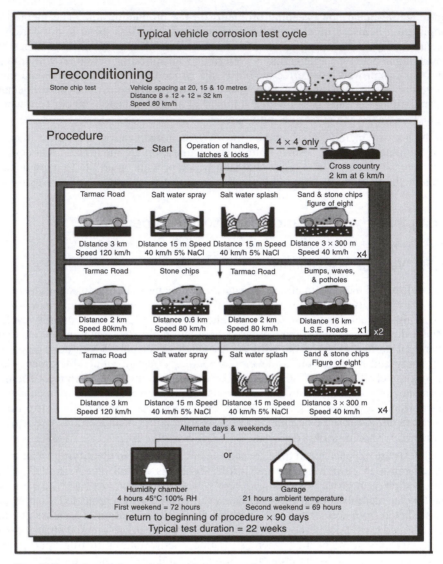

Fig. 7.8   Schematic mobile vehicle corrosion test sequence

UK, electrochemical methods are becoming increasingly useful and more relevant techniques are described below.

7.6   An introduction to electrochemical methods

The empirical tests described above allow general performance and material comparisons to be carried out and enable weaknesses to be identified in bodies or test pieces, but do not allow detailed scientific analysis to be carried out on the nature of reactions taking place at a local level. Advances in electrochemical test methods, although still relatively sophisticated and requiring delicate instrumentation, are however, becoming more robust and allowing industrial situations to be reproduced and evaluated on a more realistic scale. A brief introduction to electrochemical test methods and application may therefore be useful. The significance of these and especially the mechanistic travelling probe techniques, mapping potential and local currents, is that they can provide valuable information on simulated or real defects, with a high degree of resolution in a short time, allowing a more rapid indication of risks, and evaluation of possible protective systems. Without dwelling on thermodynamics and complex electrochemistry in detail it is still possible to appreciate the value of specific methods in assessing, the effect of spot welds for instance where these penetrate coatings, rank pretreatments and confirm the severity of damage inflicted by scratching the finished bodywork in service. For a reasonable understanding, some background knowledge in electrochemistry would be an advantage and as the emphasis is on coated steels, the galvanic nature of corrosion and protection is a good starting point.

As stated earlier, cathodic protection of the exposed steel substrate is effected by anodic dissolution of the zinc, although there also appears to be a significant barrier effect, introduced not only by the zinc itself but also by the corrosion products. However, it is the sacrificial action that dictates the choice of zinc in this context, the galvanic protection conferred on the steel substrate originating from the higher position in the galvanic series. It has long been recognized[1] that the more noble metal is preferentially protected by the sacrificial action or dissolution of the less noble coating, in this case the zinc.

The process can be compared to the reactions occurring in a galvanic cell (e.g. Cu–Zn), where a current will flow from one metal to another in the presence of an electrolyte. The anodic reaction (oxidation) will result in the release of electrons while the cathodic reaction consumes electrons to form hydroxyl ions from oxygen (reduction). For galvanized steel sheet used in automotive structures a similar situation occurs when the bodywork is scratched down to 'bare metal'. The most detrimental corrosive medium in urban environments is rock salt used for de-icing purposes during winter. When the paintwork is damaged through scratching, the salt solution (NaCl aq.) provides a suitable electrolyte, and before the advent of zinc coatings on steel, local anodes and cathodes occurring naturally on the surface would form a cell with the formation of oxides and chlorides, quickly leading to rusting and eventually perforation.

Using galvanized coatings and scratching down to the substrate offer a similar situation but now the beneficial action of the zinc can be seen as the zinc layer, being less noble than the steel, reacting sacrificially and more slowly, gives lasting protection to the steel through maintaining it at a low potential. As also indicated earlier, the corrosion products, namely, zinc hydroxides, carbonates and chlorides, can deposit on the steel and form a barrier which physically protects the substrate.

Thus a cell is initiated and this can be represented by an Evans or polarization diagram to show the relative effects of the corrosive environment on the anode and cathode. The polarization curves constructed are shown schematically in the diagrams in Fig. 7.9 extracted from an article by Kruger,[27] which illustrates the relationship between potential and current levels for both the zinc and steel surfaces. In the diagrams shown in Fig. 7.9 the most base metal (Zn) acts as the anode and the more noble (Fe) metal becomes the cathode. For zinc the open circuit potential is around −1.0 V and the value for steel is −0.44 V. Zinc has a greater effect on the steel than vice versa and can be represented by the 'cathodic' diagram shown in Fig. 7.9 and is termed as being 'under cathodic control'. The change in potential as the current increases is due to polarization of the electrode reaction[27] which changes local conditions, and which can be of three principal types ('activation', 'concentration' and 'resistance'). Polarization curves can be used to generate Tafel plots (described later), which allow the determination of important information such as the corrosion current. While potential indicates the corrosion 'tendency' of a system, it is current and current density which are a measure of the rate of corrosion. It is important to differentiate between kinetic information shown by Fig. 7.9 and the thermodynamic data conveyed by Fig. 7.10.

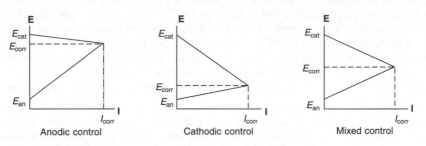

Fig. 7.9 Characteristic Evans diagrams derived from cathodic and anodic polarization curves (After Kruger[27])

Figure 7.10 shows typical 'Pourbaix diagrams' and these 'maps' illustrate the condition of zinc or steel in relation to relative pH and potential and specifically whether a tendency to passivity or reactivity exists.

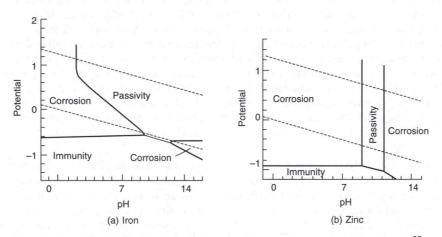

Fig. 7.10 Simplified Pourbaix diagrams for iron and zinc. (After Scully[28])

Polarization curves help to characterize the system, but to understand how specific galvanic cells, e.g. with regard to zinc coated steel, will perform the following also need to be considered:

- Relative areas of the steel and associated exposed zinc layer.
- Thickness and type of paint and pretreatments. The amount of paint applied will again determine the available area of the zinc, and the formulation of the paint, and more importantly the pretreatment, could independently influence the electrochemical behaviour of the two metals.
- Electrolyte strength. Conductivity of the electrolyte is critically linked to the distribution and size of the corrosion current and is controlled by the strength of electrolyte solution.
- Surface electrolyte thickness. The thickness of the electrolyte layer is critical in

determining the availability of reactants, e.g. access of oxygen to the metallic interface and this together with many of the above factors has been investigated by Zhang.[6] It also affects the resistance of the ionic current path.

- Exact type and composition of the substrate and protective coating (e.g. type of IZ alloy).

The composition of the electrode, as would be expected, has an influence over the corrosion characteristics and the same applies to alloy coatings, as has been demonstrated for the gamma, delta and zeta phases of iron and zinc.[13]

Considerable progress has been made recently in the use of electrochemical techniques in assessing corrosive situations. Testing of a simple nature by coupling one piece of bare metal to another (e.g. IZ to mild steel) immersed in a sodium chloride solution has been used to demonstrate fundamental differences in relative galvanic behaviour and coating longevity. Although indicating whether one metal or coating is more noble than another, this test ignores the other competing influences such as passivity, referred to above, which are readily evident from the Pourbaix diagram. To understand behaviour at small defects such as spot welds and scratches it is important that far more detailed information is also gathered which monitors local potential/current in the vicinity of local features and progressive interfacial changes taking place with time.

A comprehensive review of electrochemical methods was carried out by Sykes[29] in 1990 that emphasized the significant increase in experimental output that can now be achieved since the widespread introduction of the integrated circuit operational amplifier, digital electronics and personal computers. Together with the data acquisition equipment now available, these systems enable the operation and monitoring of complex programmes to proceed at a rate that would be difficult to achieve by manual means. Techniques relevant to the protection given by zinc coated steels at a defect include those that permit the measurement of potential, but more importantly the current density which gives corrosion rate. Cathodic and anodic polarization curves derived with computer-controlled potentiostats can be used to obtain corrosion rate from the reciprocal relationship between the corrosion current and polarization resistance (R_p) given by the Stern–Geary equation:[28]

$$R_p = \left(\frac{dE}{di}\right)_{i-0} = \frac{b_a b_c}{2.303(b_a + b_c)} \frac{1}{i_{corr}} = \frac{B}{i_{corr}} \tag{7.1}$$

where b_a and b_c are the Tafel slopes for the anodic and cathodic reactions respectively. The constant B is defined by equation (7.1). This relationship is only valid provided that both the anodic and cathodic reactions obey the Tafel equation, that is

$$n_a = b_a \log(i_a / i_{o \cdot a}) \tag{7.2}$$

$$n_c = b_c \log(i_a / i_{o \cdot c}) \tag{7.3}$$

where n_a and n_c are overpotentials and $i_{o.a}$ and $i_{o.c}$ the respective exchange current densities for anodic and cathodic reactions, but is often used in other situations with an empirical value for B.

AC impedance techniques are now in widespread use for the study of the interfacial corrosion phenomena using digital electronics to apply small AC signals of different frequencies across the interface being studied, and measuring the size and phase of the current using a frequency response analyser. Nyquist plots can be derived between real and imaginary impedance and again using computerized techniques, best fit curves can be obtained and used to define an equivalent circuit comprising capacitors, resistors and inductors. This is especially valuable for metals with organic coatings as the separate behaviour of the coating and corrosion process at defects can be determined. However, although the AC impedance methods appear attractive in helping assess actual painted situations and studying the progress of corrosive breakdown at the substrate interface, it is apparent from the work of Walter[31] that expert interpretation of Nyquist and Bode curves is needed together with relatively complex analysis. The AC method also only provides guidance on corrosion characteristics proceeding over a general area. However, as indicated earlier more detailed mechanistic techniques are now emerging from the laboratory stage of development to provide new levels of understanding on the behaviour of automotive substrate and paint systems. Based on probe analysis they involve charting the potential fields around specific features, usually defects, and deriving associated currents which indicate the associated activity. These are achieved by systematic sampling using fine probes (controlled by micromanipulators) as shown in Fig. 7.11, with sufficient resolution to examine small features, and the resultant scans are transposed to profiles or 2-D/3-D 'contour maps' for subsequent appraisal.

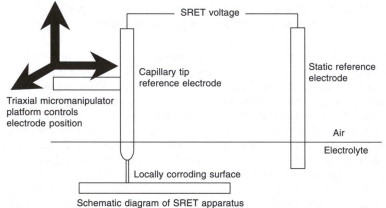

Schematic diagram of SRET apparatus

Fig. 7.11 Mapping of potential distribution using a travelling microprobe

The scanning reference electrode test (SRET),[32] which enables measurement of local potential differences, and the scanning vibrating electrode test (SVRET),[33] which measures current by measuring voltage gradients, both allow mapping over the area under study, the vibrating electrode SVRET reducing 'noise' and allowing better signal resolution. The apparent disadvantage of the SRET type of test technique, based on the principles of mapping lines of equipotential as explained by Evans and Agar,[34] and developed by Isaac[35] in the 1940s, is that the study is limited to flat metallic surfaces in contact with the relevant electrolyte. It can be used with phosphate pretreatment films and associated rinsing processes but not paint films.

A fundamental issue which must be resolved if any kind of probe/scanning technique is to be used for current derivation is that the maximum current is developed at 90° to the lines of equipotential as indicated by Evans and Agar and shown in Fig. 7.13.

Unless the lines are presented as a symmetrical matrix, any complex contour will make the derivation of corresponding true current values by intersection using a vibrating probe very difficult. Calibration in such situations is described in the review paper by McMurray[36] and the technique is especially useful for studying local activity such as cut-edges of coated steels. The effectiveness of pretreatments has also been studied using this equipment.

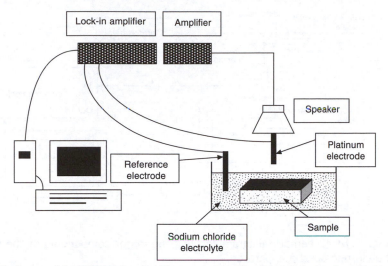

A schematic diagram of the SVRET apparatus. The sample is positioned in a tank of sodium chloride with the cut-edge under investigation horizontal and uppermost. The platinum electrode is attached to a magnetic driver (speaker) which vibrates the probe in a plane perpendicular to the sample, 125 µm above the surface, at a frequency governed by the lock-in amplifier (typically 140 Hz). Data is logged to the controlling PC.

Fig. 7.12 Scanning configuration using a vibrating probe technique to examine a coated steel cut-edge

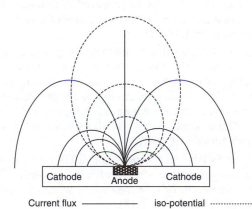

Current flux —————— iso-potential ················

Fig. 7.13 Orientation of lines of equal potential and current

The work of Stratmann[37] who uses a Kelvin probe vibrating probe technique (SKTP) shown schematically in Fig. 7.14, appears to offer a more direct method of assessing potential, with the ability to map local potential variations fairly easily and monitor activity occurring at the paint/metal interface.

The increasing attraction of SKTP is that it allows local potential measurements to be made under protective paint films, and also under thin films of electrolyte. Of particular interest in the current context is the work of Stratmann described by McMurray[36] on the collection of electrochemical data during the wet–dry cycle of atmospheric corrosion. The corrosion rate was found to increase on the drying out of the thin film of electrolyte and decrease on further drying out. This behaviour was attributed to the increasing oxygen diffusion as the film thinned and then the corrosion products

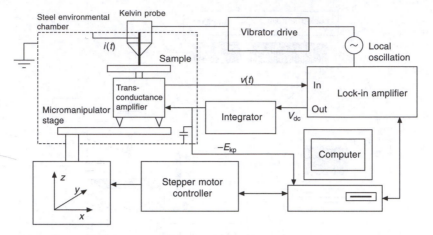

Fig. 7.14 Schematic diagram showing the major components of the scanning Kelvin microprobe apparatus

progressively passivated the surface through a blocking mechanism. Already the technique has been used to study delamination of paint from electrogalvanized substrates and to determine the effects of pretreatments on disbonding. From this work the delamination of the paint film is further explained by a galvanic cell set up between the anode at the defect/organic coating corrosion front and the more remote intact coating beneath the organic coating (cathode) of different potential. The higher the potential difference, the faster the delamination rate, and this supplements the previous theories based on hydrolysis of paint films by OH^- ions.

Recent work in this field[40] has shown some interesting results regarding automotive situations that indicate paint films of differential thickness can show unexpected results at cut-edges, the thicker film acting more anodically due to differential aeration conditions, and leading to delamination. This would confirm that holes should never be made in painted panels, especially in box sections where differential paint films are created, e.g. sill sections where a full four-coat system is applied on the outer side while inner surfaces only receive an electroprimer coat.

More recently it has been shown, using an SVRET technique,[40] that different situations can develop on opposite sides of a formed coated steel profile in IZ steel. The concave side of a semi-spherical cup appears to behave anodically whereas cracks opened on the convex side become cathodic (see Fig. 7.15).

Thus these more mechanistic techniques are already yielding some interesting results in practical situations and the sensitivity enables indications of corrosive activity to be detected at an early stage. Again with rapid advances in electronics technology it should be possible to use these quantitative techniques in a more robust mode to study on-vehicle situations under service conditions.

7.7 Learning points from Chapter 7

1. The life of automotive body structures has increased significantly during the period from 1985 to 2002 due to improved design, more effective cathodic electropainting, increased use of aluminium and the adoption of zinc coated steels.

2. Motivation for continuous improvement is fuelled by test reports provided by informed consumer organizations plus regular corrosion and strip-down tests carried

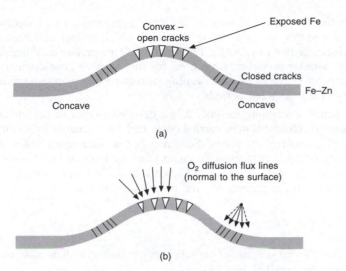

Fig. 7.15 Profiles showing different polarities developed on opposite sides of a stretch formed cup[40]

out on vehicles of market competitors. A 12-year anti-perforation warranty is now offered by leading car manufacturers on body structures.

3. Efficient design is critical in exploiting the full advantages of improved materials and processing technology. Elimination of mud traps and surface profiles prone to stone chipping are key considerations while other CAD design priorities must include optimized paint and wax access channels, e.g. castellated flange profiles and properly located drainage holes, and adequate panel separation within box sections to maximize paint coverage.

4. The quality of zinc coated steels has improved significantly during the above period allowing the levels of formability to be achieved that were previously only associated with uncoated grades. Progressive improvements in coating technology have allowed corrective rework time due to pick-up of loose zinc debris, etc. to be reduced, and most coating types now allow external paint finish standards to be obtained.

5. Although the number of different zinc coating types used by manufacturers is gradually falling, further rationalization would help the automotive industry in achieving greater product consistency and reduced costs through commonization/ interchangeability of specifications and simplification of logistics, recycling procedures, etc. This will improve through continued supplier/user technical liaison at a high level.

6. Although allowing bodyweight reduction, hybrid material combinations (e.g. aluminium + steel) can cause serious bi-metallic corrosion problems resulting in panel perforation or fastener failure when incorrect selections are made. Close reference to respective positions in the galvanic series should be made prior to specification of mixed metal combinations; manufacturing solutions to separating dissimilar materials are process dependent and vulnerable to short-range cost-cutting exercises.

7. Mixed metal bodies require special tricationic pretreatments to ensure acceptable phosphate film conversion and a consistent paint performance. Zirconium–fluoride and zirconium–titanium formulations are proving environmentally friendly alternatives to chromate conversion treatments.

8. Although automotive service conditions are difficult to reproduce on a shortened timescale, test procedures are now being specified which reflect extremes of climate more realistically, using cyclic (CCT) schedules of temperature and humidity in the laboratory. Similar procedures are used for the iterative development of vehicles, interspersing extremes of road running with overnight garage humidity exposure, over typically 22 week periods.

9. Modern electrochemical scanning methods allow detailed studies of defects and associated corrosion mechanisms to be carried out in real-time situations providing 'early warning' of corrosive situations. Scanning probe techniques allow the progressive activity occurring under automotive paint systems to be monitored very precisely at defects such as welds and rework areas in greater detail than hitherto available from more general AC and polarization techniques.

References

1. Latimer, W.M., 'The Oxidation States of the Elements and their Potentials in Aqueous Solutions', 2nd edn, Prentice-Hall, Inc., 1952.
2. Dickie, R.A., 'Surface Studies of De-adhesion', 6th Int. Conf. on Organic Coatings Science and Technology, Athens, 1980.
3. Leidheiser, H. and Wang, W., 'The Mechanism for the Cathodic Delamination of Organic Coatings from a Metal Surface', *Progress in Organic Coatings*, 11, 1983, pp. 19–40.
4. Quarshie, R.L., 'Effect of Bi-metallic Couples on Automotive Pretreatment Quality', Proc. of 6th Automotive Corrosion and Prevention Conf. SAE p-268, Dearborn, Michigan, 1993.
5. Granata, R.D., 'Mechanisms of Coating Disbondment', Proceedings of 'Automotive Corrosion and Prevention' Symposium NACE, Houston, 1991.
6. Zhang, X.G., *Corrosion and Electrochemistry of Zinc*, Plenum Press, New York, 1997.
7. Ostermiller, M.R., 'Advancements in Automotive Corrosion Resistance,' Galvatech '98 Conf., Chiba, Japan, 1998.
8. Davies, G., 'Precoated Steel: An Automotive Industry View', Metal Bulletin Monthly's International Coated Coil Conference, London, March 1993.
9. Edwards, H., Harry, E.D. and Jenkins, E., 'Galvanising: A Review of the Hot-dip Processes with some Reference to electrogalvanising', Iron and Steel Institute Symposium on Steel Strip Development, London 1961.
10. Townsend, H.E., 'Coated Steel Sheet for Corrosion Resistant Automobiles', NACE Symposium 'Automotive Corrosion and Prevention', Houston, 1991.
11. Johanssen, E. and Rendahl, B., 'A Swedish Exposure Programme of Pre-coated Steels Intended for the Motor Industry', Proc. Corrosion '91 Symposium 'Automotive Corrosion and Prevention', NACE Houston, p. 32.1.
12. Miyoshi, Y. *et al.*, 'Cosmetic Corrosion Mechanism of Zinc and Zinc Alloy Coated Steel Sheet for Automobiles', SAE Intern'l Congress and Exposition, Technical Paper No. 850007, 1986.
13. Lee, H.H. and Hiam, D., 'Corrosion Resistance of Galvanealled Steel', *Corrosion*, Vol. 45, No. 10, Oct. 1989, pp. 852–856.
14. Quantin *et al.*, 'Experience of HDG Products for Automotive Outer Panels at French Car Makers', Proc. of Galvatech '98, Chiba, Japan, pp. 583–588.
15. Berger, V., 'Concept of Corrosion Protection by Using Organic Coatings Over Galvanised Steel Sheet', Proc. Galvanised Steel Sheet Forum – Automotive, Inst. of Materials, London, 15–16 May 2000.
16. Guzman, L., Adami, M. and Voltilini, E., 'Zinc Coatings on Steel by Advanced PVD Techniques', Proc. of Galvatech '98, Chiba, Japan.

17. Schmitz, B., 'Selection and Introduction of Precoated Steel', Institute of Metals Conference on 'Moving Forward with Steel', London, 1992.
18. Technical data sheet supplied by Hoesch Steel Co., circa 1992.
19. Fujita, S., 'Corrosion Behaviour of Zinc Coated Steel Sheets on Automobile', Galvatech '98, Chiba, Japan.
20. Townsend, H.E., 'Development of Laboratory Corrosion Tests by the Automotive and Steel Industries of North America', Galvatech '98 Conference, Chiba, Japan, pp. 659–666.
21. Waddell, W. and Davies, G.M., 'Laser Welded Tailored Blanks in the Automotive Industry', *Welding and Metal Fabrication*, March 1995.
22. Suzuki, S. *et al.*, 'Edge Corrosion Resistance of Various Coated Steels', NACE Corrosion Prevention Meeting, Houston, 1991.
23. Worsley, D.A., McMurray, H.N. and Belghazi, A., 'Determination of Localised Corrosion Mechanisms using a Scanning Vibrating Reference Electrode', *Journal of Chem. Society.*
24. Strom, M. *et al.*, 'A Statistically Designed Study of Atmospheric Corrosion Simulating Automotive Field Conditions under Laboratory Conditions – Status Report on Work in Progress, Proceedings of the 5th Automotive Corrosion and Prevention Conference, P-250, Oct. 1991, paper 912282, pp. 165–178 (Society of Automotive Engineers, Inc., Warrendale).
25. Gehmaker, 'Phosphating and Chrome-free Passivation of Multi-metal Car Bodies', Proc. of Corrosion Symposium 'Automotive Corrosion and Prevention', NACE, Houston, 1991, p. 21.
26. Bucholz, K., *Automotive Engineering International*, Sept. 1998, p. 48.
27. Kruger, J., 'Corrosion Mechanisms Relevant to the Corrosion of Automobiles', Proceedings of the Corrosion Symposium 'Automotive Corrosion Control and Prevention', NACE, Houston, 1991, p. 7.1.
28. Scully, J.C., *The Fundamentals of Corrosion*, Pergamon, Oxford, 1990.
29. Sykes, J.M., '25 Years of Progress in Electrochemical Methods', *Br. Corros. J.,* 1990, Vol. 25, No. 3, pp. 175–183.
30. Stern, M. and Geary, A.L., *J. Electrochem. Soc.*, 1953, 104, 56.
31. Walter, G.W., 'A Review of Impedance Plot Methods used for Corrosion Performance Analysis of Painted Metals', *Corrosion Science*, Vol. 26, No. 9, 1986 pp. 681–703.
32. McMurray *et al.*, 'Scanning Reference Electrode Techniques Tool for Investigating Localised Corrosion Phenomena in Galvanised Steels', *Iron and Steelmaking* Vol. 23, No. 2, 1996, pp. 183–188.
33. Akid, R., 'Localised Corrosion: A New Evaluation Approach', *Materials World*, Vol. 3, No. 11, Nov. 1995, pp. 522–525.
34. Evans, U.R., *The Corrosion and Oxidation of Metals: Scientific Principles and Practical Applications*, Edward Arnold Ltd, London, 1960, p. 286.
35. Isaacs, H.S. and Vyas, B., 'Scanning Reference Electrode Techniques in Localised Corrosion', ASTM Symposium on 'Corrosion of Metals', San Francisco, Calif., 21–23 May 1979.
36. McMurray, H.N. and Worsley, D.A., 'Scanning Electrochemical Techniques for the Study of Localised Metallic Corrosion', in R.G. Compton (ed.), *Research in Chemical Kinetics*, Blackwells, 1997.
37. Stratmann, M. and Streckel, H., *Corrosion Science*, 30, 1990, pp. 697–714.
38. Duffy, L., 'Magnesium Alloys: The Light Choice for Aerospace', *Materials World*, March 1996, pp. 127–130.
39. Fontana and Greene, *Corrosion Engineering*, McGraw-Hill, New York, 1967, p. 187.
40. Private communication D. Worsley, Univ. of Swansea with G.M. Davies, Jan. 2002.
41. Ashby, M. and Jones, R.H., *Engineering Materials 2*, Butterworth-Heinemann, Oxford, 2002.

8 Environmental considerations

Objective: To review the environmental pressures relevant to the motor vehicle and consider the implications of these on the choice of body materials. In addition to well-known issues such as reduction in greenhouse gases and the importance of recycling with regard to landfill utilization, improvements in safety through additional energy absorption and more impact-friendly materials are also discussed.

Content: The wider aspects of the automotive body materials are considered: influence of the body/vehicle mass on fuel consumption and emissions – the longevity of panel materials and recyclability – importance of whole-life planning and ELV legislation – the more immediate human threats – avoidance of hazardous materials – preferred hygiene practices during assembly together with safety aspects during service and competition are also discussed.

8.1 Introduction

As an industry the automotive world has accepted its obligations with regard to the wider human issues such as the environment and safety, with foresight and considerable resources. As well as providing an increasingly efficient means of comfortable and affordable transportation it has shown a greater awareness of its responsibilities regarding preservation of both the consumer and the environment than many other industries.

The quest for lower emissions has already led to progressive improvements in controls of greenhouse and other harmful gases and developments in this direction will ultimately lead to significantly different and alternative fuels. Responsibilities for disposal of vehicles at the end of their life and funding of dismantling procedures have yet to be fully resolved but landfill limitations have been recognized and common efforts have been mounted to meet targets with regard to amounts and types of material recycled. Some form of life cycle analysis (LCA) has been adopted by most vehicle manufacturers at the earliest stage of concept design to ensure that only preferred materials meeting emissions and energy requirements – from raw material to disposal stages – are approved for production.

The responsibilities to the occupant have been demonstrated by compliance with increasingly stringent safety regulations which also require higher standards of pedestrian protection. It is therefore intended that these features, which focus on the well-being of the individual and our surroundings, are grouped together in one chapter under the heading 'Environmental Considerations' and that the major contributions made by materials developments are clearly identified.

Many of the objectives inevitably conflict, e.g. the increasing use of lightweight plastics to reduce emission levels during service raises recycling issues, and political arguments regarding the adverse consequences of weight reduction on safety in the USA retard progression on fuel economy and emission controls, but it is clear that materials will continue to play a decisive part in the overall progress made.

The contributions to lower emissions levels will be evaluated first, followed by life cycle analysis (LCA) highlighting the positive steps introduced by more progressive car companies. Finally the increasing contribution of material properties in achieving more controlled impact and collapse situations is assessed thereby meeting more demanding safety legislation. Here emphasis on energy absorbing characteristics arising from new design aspects (extrusions, TWB, etc.) and inherent developments in metallic structure.

8.2 Effect of body mass and emissions control

The choice of body materials and associated design issues was considered in Chapter 3 and brief mention was made of the rationale leading to lighter structures. As evident from work carried out on demonstrator vehicles such as the ECV3 (described in Chapter 4) the recognition of a collective automotive responsibility for, and need to, improve future fuel economy was gathering momentum in the 1970s and 1980s. According to Garrett[1] atmospheric pollution was first highlighted in Los Angeles in 1947, a Dr A.J. Haagen-Smit claiming through his research that this was mainly due to automotive exhaust emissions.

The harmful effects of exhaust gases can be understood from the following explanation.

Assuming complete combustion,[1] each kilogram of hydrocarbon fuel when completely burnt produces mainly 3.1 kg of CO_2 and 1.3 kg of H_2O. Most of the undesirable exhaust emissions are produced in minute quantities and these are oxides of nitrogen NO_X, unburnt hydrocarbons (HC), carbon monoxide (CO), lead salts, polyaromatics, soots, aldehydes, ketones and nitro-olefins. Since the 1980s the concern over CO_2 has mounted, not due to its toxicity but because it was suspected of facilitating the penetration of the atmosphere by ultraviolet rays emitted by the sun. Carbon monoxide is toxic because it is absorbed by the red corpuscles of the blood, inhibiting absorption of the oxygen necessary for sustaining life. The toxicity of hydrocarbons and oxides of nitrogen arise through their photochemical reactions with sunlight leading to the production of other chemicals. There are two nitrides of oxygen, NO and NO_2, and under the influence of solar radiation the NO_2 breaks down to NO + O and the highly reactive oxygen then combining with O_2 to form ozone, O_3. The presence of hydrocarbons inhibits the reaction whereby it recombines with NO and reverts to NO_2 and the concentration of ozone rises. The ozone then proceeds to form chemicals which combine with moisture to form smog.

This prompted early legislation to stem urban pollution, catalytic converters providing an interim solution. Later wider concerns were to develop surrounding greenhouse gases. In summary the greenhouse effect refers to a natural effect whereby a layer of gas of industrial origin is built up, causing infrared radiation to be trapped in the atmosphere (as shown in Fig. 8.1), thus warming the surface of the earth and causing climatic changes leading to well-known speculation on a number of different subjects. This process is explained more fully in the text *Global Warming* by J. Legget.[2]

Although by no means the only protagonists, the automotive industry was regarded as major contributors to pollution effects. Coincidentally in 1973 the Arab oil embargo and quadrupling of the OPEC oil prices led to American Congress introducing Corporate Average Fuel Economy (CAFÉ) legislation in the USA. This required that for all cars marketed by each corporation selling to the USA the consumption had to improve in stages from 18 mpg in 1978, then by 1 mpg each year to 1980. Thereafter 2 mpg annually to 1983, again by 1 mpg for 1984 and 0.5 mpg to 27.5 mpg for 1985. Since then environmental lobbyists have attempted to raise this figure to 40 mpg but this has been resisted by the safety lobby who claim a link between the frequency of fatalities and downsizing of vehicles to improve fuel economy.

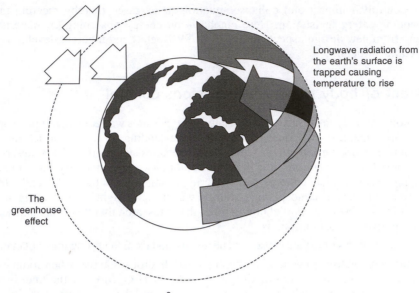

Longwave radiation from the earth's surface is trapped causing temperature to rise

The greenhouse effect

Fig. 8.1 Greenhouse effect[2]

Bodyweight reduction has often appeared at variance with the addition of other vehicle components where safety and even other methods of emissions control have progressively added weight through items such as air bag equipment, more sturdy front/rear energy absorbing armature designs and catalytic converters, etc. PSA Peugeot Citroen[3] in fact recorded a 150 kg average increase in vehicle weight from 1985 to 1995. Additionally the add-on options, including enhanced audio systems, trip computer and navigational aids, have all become more popular in response to the increasing consumer expectations and competitive initiatives, and again the aim of the PSA group is to reduce weight by 5 to 15 per cent by 2008 to offset these items, utilizing lighter materials. The body designer has generally managed through parts redesign, parts consolidation and sensible materials substitution to deliver a steady weight reduction progression over the last 30 years. This is indicated over three generations of BMW models shown in model generation/body to kerbweight diagrams (Fig. 2.14) referred to in Chapter 2 and this is expected to continue into the future.

To put the potential savings available from bodyweight reduction in perspective with other design parameters affecting fuel economy, a 10 per cent power train efficiency improvement yields a 10 per cent improvement in fuel economy. Vehicle weight yields a 5 per cent improved fuel economy while 10 per cent difference in rolling resistance (tyre friction) effects a 1.3 per cent reduction. Aerodynamics and styling is more dependent on driving conditions, a 10 per cent reduction in wind resistance under rural conditions giving a 3 per cent improvement in fuel economy whereas the same 10 per cent would yield 5 per cent economy under motorway driving conditions (120 km/hr). For a test cycle used to assess CO_2 emissions, this would fall to 1.5 per cent. Therefore the contribution of bodyweight (around 30 per cent of the vehicle weight) reduction remains highly significant and various estimates have been quoted for the specific savings achieved by lightening this structure. Peugeot[3] claims a 50 kg reduction in vehicle weight induces fuel savings corresponding to a reduction in CO_2 emissions of 2 to 3.5 g/km. The momentum of achieving better levels of performance has also been maintained by competitive developments with one manufacturer keenly aware of the capability of market rivals.

The number and type of vehicles produced by a manufacturer obviously affect their corporate performance, but in Europe it is predicted that most will achieve 6 l/100 km by 2005, although those with a bias towards performance cars might only achieve a figure in the region of 7 l/100 km. A key milestone for manufacturers is a volume vehicle which will achieve 3 l/100 km, and already this has proved attainable for a number of Japanese and European manufacturers including VW, Volvo, Ford, Mercedes, Audi, Opel and Toyota.

As well as the EPA (Environmental Protection Agency) monitored CAFÉ regulations in the USA similar directives are also evident in Europe where tax incentives now exist in Italy, Germany and the UK – as evident from the £55 rebate in annual road tax applicable to vehicles under 1400 cc capacity. The industry has to agree to monitoring by 2003, and achievement of the EU voluntary level of 140 g/km carbon dioxide by 2008.

PSA share this objective and recent figures[3] suggest this is attainable:

* In 1998 the group ranked fourth in Europe with 175 g/km of CO_2, 9 g behind the leader
* In 1999 this level was reduced to 168 g/km
* In 2000 the indications were more favourable at 162.7 g/km of CO_2

In mid-2002 the top-ten performing cars in Great Britain registered values between 80 and 119 g/km. In the UK tax on company cars from 6 April 2002 has changed from being mileage related to being based on CO_2 emissions and a percentage of the list price. This initiative is a key factor in the government's Climate Change Programme.

The link with materials choice and the obvious candidate materials have already been referred to but it is clear from the above figures that manufacturers are adopting an extremely positive attitude with regard to their environmental responsibilities as will further become clear with the other issues outlined below. The major European steel companies hold regular seminars with the automotive industry to discuss their needs and EU Technical Programmes are promoted regularly to further develop metallic and non-metallic weight reduction initiatives. Despite the emphasis on steel and aluminium and problems evident from attempts to integrate them for body parts in large volume projects, plastics are very attractive for smaller volume niche markets. This latter point is reinforced by a recent press report[4] which highlights the particular attraction of the plastic bodied Lotus Elise, in that weighing only 750 kg it will attract a lower rate of taxable benefit due to a more favourable level of CO_2 emissions than other market contenders with a similar high performance. The same applies to other sports cars.

Although the same benefits for weight efficient designs referred to above will apply for other types of fuel (i.e. lighter body less fuel consumption) it will be interesting to observe how policies might change as other fuels such as hydrogen and electricity become the norm and CO_2 emissions become less of an issue. Serious initiatives are already well advanced by the major vehicle manufacturers such as BMW who have adapted 7 series saloons and even the Mini for demonstrating that today's internal combustion engines can be converted relatively easily to run on hydrogen with water vapour as the major gaseous emission. Reports[5] suggest that the UK government would encourage 10 per cent of all new cars to run on hydrogen instead of fossil fuels within a decade. This would result in 200 000 cars a year with no pollution, going some way to satisfying Britain's obligations under the Kyoto environmental agreement to reduce carbon dioxide output. Hydrogen appears to be the longer-term solution as practicalities such as a supporting network of hydrogen filling stations would have to be put in place. Hybrid systems such as available on the Toyota Prius may provide an intermediate stepping stone. These feature a petrol engine/electric motor hybrid, and at low speeds the Prius petrol engine shuts down and the electric motor takes over and hence in heavy traffic it is emission free. Overall CO_2 emissions are reported as 114 g/km. The next step in this development chain leads to fuel cells whereby electric current is generated by combining hydrogen and oxygen using a catalytic converter. This technology is outside the scope of this chapter but it may call for the re-examination of materials strategy if the gases referred to above are replaced by relatively safe alternatives such as water vapour. If the offending gases are not generated – even by structures twice as heavy – is there still a need to adopt alternative materials? Is there still a need for CAFÉ regulations if alternatives to oil are available? In the interim period while fuel tanks and generating equipment are still in the 'cumbersome' stage for alternative fuels, accompanying structures must be of minimum weight, but as long as these are downsized it is likely that the materials which meet the usual process chain and safety criteria most easily, i.e. steel and aluminium, will be selected. Plastics obviously offer a lot of potential but as will be seen later in this chapter a solution must be found for the recycling problem.

So far only the 'in-service' aspects of emissions generation and control have been considered but raw material manufacture and end-of-life (ELV) disposal also involve an energy input, and increasing numbers of automotive producers are now adopting a cradle-to-grave approach with all implications considered before a material selection is made. The next section illustrates how all aspects of the life cycle are considered.

8.3 Life cycle analysis (LCA)

Throughout the 1990s larger corporations had been urged to take a more environmentally responsible attitude to material selection and this prompted a number of complex proposals for assessing the magnitude of the different impacts at various stages of manufacture, life cycle analysis. As the term suggests an evaluation is made of the impact of one effect attributable to a material, e.g. CO_2 emissions, not simply during the period the vehicle is in service but through the overall lifespan from material manufacture to forming, BIW assembly and recycling.

Criteria can differ widely and in addition to CO_2 emissions, the focus could be energy or costs, etc. The complexity of LCA arises through the methodology selected although the lack of uniformity has been addressed by groups such as the Society of Environmental Toxicology and Chemisty (SETAC) and the International Organization for Standardization (ISO). An appreciation of the complexity of some of the issues involved can be gained from recent review publications[6] but so far as the current materials choices are concerned two types of audit are presented for steel and aluminium, the first compiled in detail from widely gathered empirical data at each stage. The other used a more practical-based programme where an aluminium-based structure was compared against the market norm.

Audits for energy, CO_2 emissions and cost have already been derived for steel and aluminium by the IISI[7] and summaries extracted from their 1994 report, starting with the smelting and refinement process (material production phase), followed by the process chain production sequence (BIW transformation phase), the in-service stage (utilization phase) and concluding with the disposal (recycling of used car phase) are presented in Tables 8.1 to 8.3. Some bias might be expected in the derivation and interpretation of some of the data due to the allegiance of the research source to ferrous materials, but an attempt has been made to present values showing most favourable and unfavourable cases for both steel and aluminium.

Table 8.1   Compared energy balance of steel and aluminium BIW

	Steel favourable scenario			Aluminium favourable scenario		
	Present steel BIW	Future steel BIW	Potential Al BIW	Present steel BIW	Future steel BIW	Potential Al BIW
1. Material production phase						
BIW net weight in kg	280	224	182	280	224	168
BIW raw weight in kg	407	326	273	407	326	252
Material production MJ/kg	30	30	170	30	30	170
Material production MJ/BIW	12 222	9778	46 410	12 222	9778	42 840
Material recycling MJ/kg	12	12	18	12	12	18
Economy on primary metal production through recycling of stamping scrap MJ/BIW	−2247	−1798	−13 279	−2247	−1798	−12 257
Material transportation to car plant MJ/BIW	7	6	14	7	6	14
Total material production phase	9982	7986	33 145	9982	7986	30 597
2. BIW transformation phase						
Stamping	1019	1019	1324	1019	1019	1324
Joining	1050	1050	3150	1050	1050	3150
BIW tempering	0	0	1300	0	0	1300
Total BIW transformation phase	2069	2069	5774	2069	2069	5774
1 + 2 BIW production phases	12 050	10 054	38 919	12 050	10 054	36 371
3. Utilization phase (10 years)						
Fuel consumption 1	2250	2095	1980	3000	2525	2050
Fuel consumption MJ	72 562	67 578	63 840	96 750	79 412	70 744
Fuel production MJ	8935	8321	7861	11 914	10 028	8142
Total utilization phase MJ	81 498	75 899	71 701	108 664	89 440	78 886
1 + 2 + 3 BIW production and utilization phases	93 548	85 954	110 620	120 714	99 494	115 256
4. Recycling of used car phase						
Production of liquid metal from scrap MJ/kg	5	5	12	5	5	12
Production of liquid metal from ore MJ/kg	22.5	22.5	160	22.5	22.5	160
Actualization factor	0.5	0.5	0.5	1.0	1.0	1.0
Economy of primary metal production through material recycling from used car MJ/car	−2121	−1697	−10 717	−4243	−3394	−19 784
Total energy balance of a BIW MJ	91 427	84 257	99 903	116 471	96 100	95 472

The conclusions of this report suggest:

- Legislators need to be more widely informed of true energy and ecological consequences when encouraging work on lower density metallic materials.
- Potential purchasers should seek proof of claimed advantages for lighter bodied vehicles. In the case of aluminium intensive vehicles any resultant fuel savings should be carefully balanced against negative factors arising from higher repair costs, delays due to the limited number of authorized repair agents and the possibility of higher insurance category ratings in the future.

To counter this evidence from the steel sector a further study is quoted by Peters[8]

Table 8.2   Compared CO_2 balance of steel and aluminium BIW

	Steel favourable scenario			Aluminium favourable scenario		
	Present steel BIW	Future steel BIW	Potential Al BIW	Present steel BIW	Future steel BIW	Potential Al BIW
1. Material production phase						
BIW net weight in kg	280	224	182	280	224	168
BIW raw weight in kg	407	326	273	407	326	252
Material production kg CO_2	1.8	1.8	10.5	1.8	1.8	10.5
Material production kg CO_2/BIW	733	587	2867	733	587	2646
Material recycling kg CO_2	0.40	0.40	0.66	0.40	0.40	0.66
Economy on primary metal production through recycling of stamping scrap kg CO_2/BIW	−175	−140	−860	−175	−140	−793
Total material production phase	559	447	2007	559	447	1853
2. BIW transformation phase						
Stamping and joining	103	103	224	103	103	224
BIW tempering	0	0		0	0	
Total BIW transformation phase kg CO_2/BIW	103	103	224	103	103	224
1 + 2 BIW production phases	662	550	2231	662	550	2076
3. Utilization phase (10 years)						
Fuel consumption 1	2250	2095	1980	3000	2525	2050
Fuel consumption kg CO_2	5243	4882	4612	6990	5884	4777
Fuel production kg CO_2	526	489	462	701	590	479
Total utilization phase kg CO_2/BIW	5768	5372	5075	7691	6473	5256
1 + 2 + 3 BIW production and utilization phases kg CO_2	6430	5922	7305	8353	7024	7332
4. Recycling of used car phase						
Production of liquid metal from scrap kg CO_2	0.25	0.25	0.6	0.25	0.25	0.6
Production of liquid metal from ore kg CO_2	1.5	1.5	10	1.5	1.5	10
Actualization factor	0.5	0.5	0.5	1.0	1.0	1.0
Economy of primary metal production through material recycling from used car kg CO_2/car	−152	−121	−681	−303	−242	−1257
Total CO_2 balance of a BIW kg CO_2	6279	5801	6625	8050	6781	6076

based on the ACCESS initiative, an aluminium-based concept car with fully recyclable aluminium spaceframe, plastic outer panels and aluminium/plastic laminate for the roof and bonnet. The environmental impact compared with model year cars in a similar compact size category are shown in Fig. 8.2.

To help assess the effect of polymer composites reference is made to the 'Hypercar'[9] referred to in Chapter 4 which shows a predominance of plastic and composites (43 per cent) compared with metals of only 42 per cent. It is claimed that the Hypercar would weigh only one-third as much as the steel car it would replace yet the embodied energy would not be three times higher per kilogram. Typical values quoted are 77–121 MJ/kg for most polymers (carbon fibre not being exceptional), 342 MJ/kg for

Table 8.3   Compared cost balance of steel and aluminium BIW vehicles

	Steel favourable scenario			Aluminium favourable scenario		
	Present steel BIW	Future steel BIW	Potential Al BIW	Present steel BIW	Future steel BIW	Potential Al BIW
1. Material production phase						
BIW net weight in kg	280	224	196	280	224	168
BIW raw weight in kg	407	326	294	407	326	252
Material cost $/kg	0.68	0.7	3.81	0.68	0.7	3.81
Material cost $/BIW	227	228	1120	227	228	960
Scrap value $/kg	0.08	0.08	1.10	0.08	0.08	1.10
Scrap value $/BIW	−10	−8	−108	−10	−8	−92
Material transportation to car plant and storage costs $/BIW	16	13	16	16	13	16
Total material production phase	283	233	1028	283	233	884
2. BIW transformation phase						
Stamping	200	200	300	200	200	300
Joining	250	250	450	250	250	450
Specific tooling	350	350	420	350	350	420
BIW tempering	0	0	30	0	0	30
Painting	300	300	450	300	300	450
Total BIW transformation phase	1100	1100	1650	1100	1100	1650
1 + 2 BIW production phases	1383	1333	2678	1383	1333	2534
3. Vehicle price						
Equipment and assembling	3600	3600	3600	3600	3600	3600
Fixed costs	1300	1300	1300	1300	1300	1300
Vehicle production cost	6566	6466	8607	6566	6466	8317
Car maker and dealer margin − VAT	3073	3026	4028	3073	3026	3892
Vehicle price	9639	9492	12 634	9639	9492	12 210
4. Utilization phase (10 years)						
Fuel consumption 1	9000	8791	8687	12 000	11 335	10 670
Fuel cost for vehicle life $	9000	8791	8687	12 000	11 335	10 670
Interest for capital investment	5783	5695	7581	5783	5695	7326
Insurance cost	7000	7000	8750	7000	7000	8750
Total utilization phase cost $	21 783	21 487	25 018	24 783	24 030	26 746
Vehicle price and utilization cost	31 423	30 979	37 652	34 423	33 522	38 956
5. Recycling of used car phase						
Scrap value	100	100	800	100	100	800
End of use of vehicle value $	60	60	169	60	60	149
Total cost balance of a vehicle $	31 363	30 919	37 483	34 363	33 462	38 807

aluminium and 64–129 MJ/kg for steel. It would be difficult to make a full LCA comparison as no figures exist for fabrication and recycling. However, reference is made to the Renault Espace where manufacturing processes for polymer body panels are by far less capital intensive than for steel or aluminium and tooling cycles much faster. The drawback for all plastics however, as will be seen in section 8.3, is recycling – a major life cycle issue. As pointed out by Price[10] 'thermoplastics must be ground or broken down into small particles for re-melting and remoulding. This process could destroy the original reinforcement fibre length and strength. Thermoset-

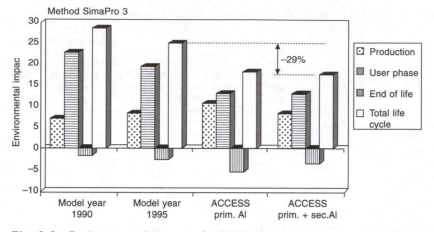

Fig. 8.2 Environmental impact of ACCESS aluminium structured vehicle. (Courtesy SAE[8])

based composites must be broken down into particles of varying sizes but can then only be employed as an inert filler. These are environmental drawbacks which become costs and are passed on to the end users. So composites face three re-cycling challenges:

(1) how to identify some 200 kinds of polymers that can be found in the average vehicle

(2) how to separate them

(3) how to develop new products with a Class "A" finish from the re-cycled materials.'

The 200 types may apply to the whole vehicle and another report[11] refers to 20 different polymer types used in automotives (PP, PUR, ABS, PE, PVC and PA) although six of these account for 70 per cent of applications.

All these require different preparation, and as we shall see later, disposal routes and comparative data are more limited than the metals shown above. Regarding the transformation phase, however, it is generally accepted that, as in Fig. 2.37, p. 52, the production costs are lower at lower volumes for plastic panel production, confirming the Espace experience,[9] a major factor being the cheaper tooling used.

Thus the methods used in LCA comparisons can be appreciated and must be applied to body panels in the future, although due to complexity and arguments on methodology, uniformity is still a key requirement. Most large corporations carry out this type of audit as an early part of new concept design. The responsible attitude taken by automakers is illustrated by the procedure adopted by BMW whereby no new material is considered for use unless an audit of whole life implications of the material is understood and approved by experts.

8.4 Recycling and ELV considerations

'ELV' ranks with CO_2 emissions as being the most well known of the automotive environmental topics and refers to the subject of end-of-life vehicles recycling which despite improvements in various aspects of disposal technology, leaves 25 per cent of the vehicle weight going to landfill. Although this is mainly non-metallic plastic, rubber and glass, and the scope refers to the whole vehicle body, materials feature prominently and a coherent strategy must exist for their selection and, ultimately,

1

The last owner delivers his ELV to an authorized recycling company. The recycling process starts.

2

Details of every vehicle that is accepted are recorded and classified according to type. The model and the vehicle's condition determine the processing methods that have to be used, and also the residual value.

3

The first stage in the actual process is to remove all operating fluids and similar substances: oils, fuel, air-conditioning refrigerant, coolant and brake fluid.

4

If engines are stripped down expertly in a high-value recycling process, they can be reconditioned and re-used with no loss of quality. They serve as input material for BMW's rebuild part production plant in the south-east German town of Landshut.

Fig. 8.3 Whole life approach adopted by a major manufacturer. (Courtesy of BMW Group[14])

6

Glass and many plastics can already be recovered economically by today's methods. The BMW Group has contributed greatly to this by developing suitable new methods.

9

According to the EU directive, only 5 per cent in weight of the residual material may be dumped from 2015 on; this is approximately the same amount as would fit into a typical 80-litre domestic garbage bin.

8

The shredder cuts up the compressed body-shell blocks into pieces no larger than the palm of the hand. Ferrous and non-ferrous metals are separated for further processing.

5

According to the EU directive, 85 of every 100 kilograms of the end-of-life vehicle's weights must be re-used, either as parts or as materials. This require-ment has long since been fulfilled as far as metal parts are concerned.

7

What remains of the vehicle is compressed into a solid block of metal; this reduces the cost of transportation to the shredder.

Fig. 8.3 (Contd)

reuse. To re-enforce the importance of this topic initiatives are first described in Europe and the UK before illustrating moves in various other parts of the world, again using one manufacturer's general management of the subject as an example.

8.4.1 The European recycling programme

The central feature of the European initiative is the EU End of Life (ELV) Directive (2000/53/EC) which came into force on 21 October 2000. All member states were due to transpose the Directive into national law by 21 April 2002 although this deadline has been missed by a number of countries.

A DTI Consultation Document issued in August 2001 summarizes the situation thus:

> The overall objective of the ELV is to reduce the amount of waste generated during the scrapping of vehicles. In particular it:
>
> - requires member States to ensure that ELVs can only be scrapped by authorized dismantlers or shredders, who must meet tightened environmental treatment standards from the outset;
> - requires economic operators (this term includes producers, dismantlers and shredders among others) to establish adequate systems for the collection of ELVs from the outset;
> - states that last-owners must be able to return their vehicles into these systems free-of-charge from January 2007;
> - requires producers (vehicle manufacturers or importers) to pay 'all or a significant part' of the costs of takeback and treatment from January 2007. Member States can also apply this requirement from the outset.
> - sets rising re-use and recovery targets (re-cycling targets) which must be met by economic operators by January 2006 and 2015;
>
> (no later than 1.1.2006, re-use and recovery for all ELVs must be 85%, of which a minimum of 80% must be re-cycled,
>
> no later than 1.1.2015, re-use and recovery for all ELVs must be 95%, of which a minimum of 85% must be re-cycled) and
>
> restricts the use of heavy metals in new vehicles from July 2003.

The Directive aims to introduce a coherent plan to control recycling effectively across European member countries with benefits in shared planning and co-operation in implementation.

Under the terms of the Directive there is an obligation on the manufacturer to ensure the final destruction of a vehicle in the most environmentally friendly way and it applies to passenger carrying vehicles and light goods vehicles of 3500 kg or less. Organizations producing less than 500 units per year are exempt. Mass produced vehicles must be delivered to authorized treatment facilities (ATF). A Certificate of Destruction (CoD) will need to be issued to the owner on final receipt at the ATF to inform the DVLA (UK Licensing Authority) of the vehicle's obsolescence.

As shown in Table 8.4 progress has already been made in the UK in recent years to increase the amount of material recycled.

The source of this data was ACORD, a consortium of interested organizations formed in 1997 when the Society of Motor Manufacturers and Traders (SMMT), the British Metals Federation (BMF), the Motor Vehicle Dismantlers Association (MVDA), the

Table 8.4 ELV performance summary 1997–99. (Source: ACORD 2001)[12]

Vehicles scrapped – (units)	1 800 000	100	1 800 000	100	2 017 000	100
Total material for disposal – (tonnes)	1 884 000	100	1 884 000	100	2 108 000	100
Parts reused – (tonnes)	193 000	10%	193 000	10%	240 000	11%
Materials recycled – (tonnes)	1 205 500	64%	1 253 500	67%	1 460 000	69%
Total recovery – (tonnes)	1 398 500	**74%**	1 446 500	**77%**	1 700 000	**80%**

British Plastics Federation (BPF) and the British Rubber Manufacturers Association (BRMA) signed an agreement regarding the handling of ELVs and pledged to improve recovery and recycling measures in the UK.

The existence of such an organization underlines the complexity of the chain involved in disposal, a more detailed analysis being shown in Fig. 8.4.

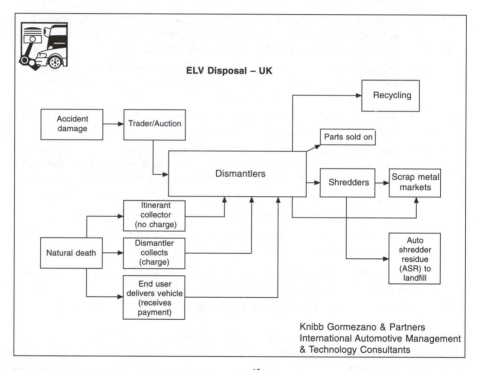

Fig. 8.4 Schematic of ELV disposal routes[13]

A further organization, CARE (Consortium for Automotive REcycling in the UK) comprises the key vehicle manufacturers including BMW (GB) Ltd, Fiat, Ford/Jaguar, Mercedes-Benz (UK) Ltd, PSA (Peugeot/Citroen), Renault, VAG (UK) Ltd, Vauxhall, Volvo and Rover together with dismantlers. Among other advantages the dismantlers have ready access to the IDIS electronic dismantling information system which assists in the identification and segregation of the various plastics used by manufacturers. All participating car manufacturers are committed to a policy of specifying recycled materials hopefully leading to the increased demand and improved economics in the future. Materials specifically targeted for recovery include polyester, ABS and polypropylene. Seventy-five per cent of the vehicle weight is currently recycled and the remainder comprises mainly the materials mix known as auto shredder residue (ASR) or 'fluff' which normally goes to landfill. With between 9 and 13 million vehicles being scrapped in Western Europe every year it has been estimated that fluff constitutes 10 per cent of the hazardous waste generated in the EU annually. In the UK 1.8 million vehicles every year are scrapped (out of a total of 23 million) representing 500 000 tonnes of ASR being dumped in landfill. There are about 3000–4000 dismantlers operating in the UK who feed 30–40 shredders. Once dismantled for parts reuse, the remaining ELV hulk is usually compressed (with or without the engine) for transportation to a shredder for metal recovery. Aluminium body panels of the few specialist cars can either be recovered by disassembly or through post shredder separation techniques (no major influx of aluminium bodied cars into dismantlers yet).

Steel and aluminium body panels are relatively easily recovered as the body shell is disassembled and constitute a high proportion of the vehicle metallic content. The total content can be broken down as shown in Table 8.5.

Table 8.5 Material breakdown for passenger cars[12]

	1997	1998/9/0	1997	1998/9/0
Ferrous metal	68.6	68.3	773	780
Light non-ferrous	6.1	6.3	68	72
Heavy non-ferrous	1.8	1.5	20	17
Electrical/Electronics	0.7	0.7	8	8
Fluids	2.1	2.1	23	24
Plastics	8.5	9.1	96	104
Carpet/NVH	0.6	0.4	6	4
Process polymers	1.2	1.1	14	12
Tyres	3.5	3.5	40	40
Rubber	1.7	1.6	19	18
Glass	2.9	2.9	33	33
Battery	1.1	1.1	13	13
Other	1.2	1.5	13	17
Total[(a)]	100	100	1126	1142

(a) Passenger cars only. Average van weight is around 1480 kg.

The higher the scrap value of steel, aluminium and other non-ferrous metals, the more the incentive for recovery and while almost 50 per cent of the prime cost of aluminium alloys can be recovered (if segregated), the cost of steel bales remains low (nearer 10 per cent). However, both industries have fully active recovery programmes extending to reutilization of zinc dust from steel furnaces[14] when galvanized sheet is recycled. As already stated, plastics are more difficult to reuse partly because of the number of derivatives, although initiatives are evident, as referred to when describing the ELV status in the USA, below.

8.4.2 A manufacturer's policy

Most car manufacturers have already installed in-house procedures to anticipate legislation and this extends to established relationships built up with suppliers, motoring research bodies, insurance and repair organizations, the recovery industry, and after-sales networks to ensure a co-ordinated approach is adopted by the industry.

Measures adopted generally include:

- Incorporation of a marking system to identify different plastics in product design standards
- Adoption of procedures to ensure easier dismantling of parts

- Ensuring compliance with hazardous substance regulations
- Adoption of electronic identification system IDIS, increasingly adopted by the industry for segregation of plastic and other parts
- Participation in discussion with relevant organizations, e.g. ACORD Voluntary Intersector Agreement on ELVs 1997, CARE automotive group
- Installation of dismantling research centres where strip-down procedures can be carefully planned for existing and future models

From the references made above and the next section on worldwide policy it is evident that all the major vehicle manufacturers already have corporate systems in place to ensure an 'ecologically optimized' approach is adopted both within their in-house operations and supply chain.

Consultation with national legislative bodies and competitor/associated groups with similar interests maintains an awareness of timescales and provides a balanced feedback to government of any realistic inability to meet deadlines or apportionment of responsibilities. This is bound to happen if the ELV treatment and recycling costs cannot be absorbed by the recovery industry, and payment by last owner must move to shared responsibility by producers and others from 2007.

The joint initiatives being taken in dealing with plastics with suppliers and the headway in establishing preferred groups of materials that will be more easily treated in the future show the lateral co-operation that will be necessary to simplify the position and if increasingly demanding targets are to be met in the future.

Any prominent car manufacturer should have a clearly identified internal policy that can be used as a reference throughout the organization and can be implemented at the earliest possible design stage and followed through to the ELV stage. The system adopted by BMW in this respect is extremely detailed and has been very carefully researched and implemented within the company since 1990. The company is also extremely aware of all ecological aspects and a life cycle analysis is carried out on all materials even considered at the concept stage. The following passages are reproduced from literature issued by BMW and although much of the content covers the whole vehicle the principles are the same when applied to the body materials. The key points are summarized as follows and emphasize foresight in materials selection, Fig. 8.5(a), (b) and (c) conveying different aspects of the BMW brochure.

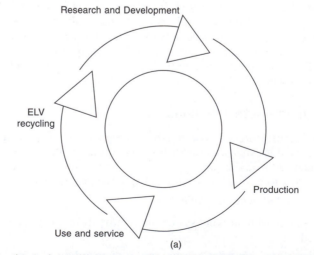

Fig. 8.5 (a), (b) and (c) Illustration of key initiatives taken by BMW[15]

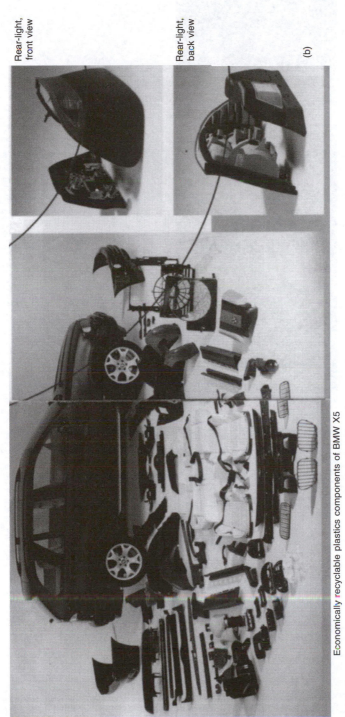

Rear-light,
front view

Rear-light,
back view

(b)

Economically recyclable plastics components of BMW X5

Fig. 8.5 (*Contd*)

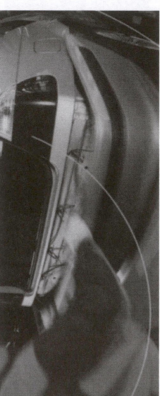

Dismantling rear bumper

(c)

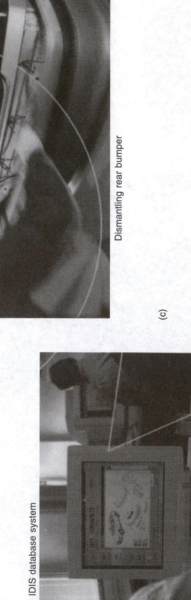

IDIS database system

Fig. 8.5 *(Contd)*

8.4.3 Other progress worldwide

A recent state-of-the-art report[16] describes initiatives in the USA. First, in the 10th biennial University of Michigan Delphi Forecast and Analysis it is predicted that total vehicle weight will decrease 10 per cent for both passenger cars and light trucks by 2009 in part due to increases in CAFÉ requirements but in the view of the 200 experts consulted the recyclability of automotive materials and related concerns will pose significant challenges again referring to 'about 75 per cent (chiefly metallic content) presently recycled and 25 per cent ending up in landfills'. This was regarded as an important statement as the introduction of lighter materials will almost certainly clash with ELV expectations. Reference is made to instances of corporate policies adopted by the three major automotive companies – although many other examples could be highlighted.

Ford are carrying out major initiatives on plastics recycling and have announced a power-train application using repolymerized nylon with the Visteon Corporation to make a high performance throttle-body adaptor, from recycled material including carpeting. In Europe co-operation has been requested by Ford from all suppliers for their Corporate Vehicle Recycling Strategy.

The OEMs are quoted as taking a proactive stance and the views of the Saturn organization are particularly relevant as they represent a company with a significant number of body panels in plastic, supported by a metallic substructure. A feature of future policy will be on strengthening partnerships with the Saturn 400 suppliers on environmental issues with as strong a reference to ISO 14001 as there has been on the ISO 9000 quality standard. Reprocessing and recycling have been designed into the Saturn manufacturing process with more than 35 per cent of each vehicle made from recycled material including steel from the spaceframe, aluminium in engine and wheels and reprocessed polymers for body panels. Teaming up with General Motors Research and a group of suppliers Saturn have also developed a breakthrough for painted plastics recycling with use for panels like wheel arch liners and rocker panels following regrinding.

Daimler-Chrysler have a further plastics recycling initiative which represents the second phase of their CARE (Concepts for Advanced Recycling and Environmental) car programme and help the company achieve its goal of vehicles that are 95 per cent recoverable by 2005. Using a skin flotation process it appears feasible to separate various plastics and rubbers which is the first step to their reuse.

In Japan the lack of land for in-fill sites is critical and initiatives began in 1997 when MITI (Ministry of International Trade and Industry) launched an ELV initiative. The Japanese Automobile Manufacturers Association (JAMA) then established its own voluntary action plan which extends to a number of industry and external partnerships. The rate of recycling has risen from 75 per cent to nearer 80 per cent in weight in recent years. Improvements have been made at the dismantling stage and there is a standardized materials identification system. The emphasis has been on co-operation between all industry sectors with a checklist system in place to ensure compliance with appropriate ELV procedures throughout all sectors involved.[17]

8.5 Hygiene

Although it is apparent that the automotive industry has for some time operated its own form of voluntary hygiene controls and worked to strict TLV levels* and codes

---

*Threshold limit values are used as a measure of toxicity. The TLV is the maximum ambient air concentration to which a worker can be exposed without adverse effect (mg/m^3 at 8 hours' exposure day or 40 hours' work-week).

of practice for working with hazardous substances, further emphasis must be placed on such controls at the disposal stage.

A further stipulation of Directive 2000/53/EC is that 'preventative measures be applied from the conception phase of the vehicle onwards and take the form, in particular, of reduction and control of hazardous substances in vehicles, in order to prevent their release into the environment, to facilitate recycling and to avoid the disposal of hazardous waste. In particular the use of lead, mercury, cadmium and hexavalent chromium should be prohibited. These heavy metals should only be used in certain applications according to a list which will be regularly reviewed. This will help to ensure that certain materials and components do not become shredder residues, and are not incinerated or disposed of in landfills.'

This introduces a further aspect of body material technology – hygiene control and this applies to the selection, application and disposal of commonly used materials and related treatments.

8.5.1 Heavy metal restrictions

Materials and components of vehicles put on the market from July 2003 must not contain the heavy metals referred to in the opening paragraph other than applications listed in Annex 10 of the Directive:

- Free cutting steels containing 0.35 per cent lead
- Aluminium alloys containing 0.4 per cent lead
- Lead bronze engine bearings
- Lead-acid batteries
- Fuel tank lead coatings
- Lead vulcanizing agent for hoses
- Lead stabilizer in paints
- Lead solder
- Hexavalent chromium up to 2 g per vehicle
- Mercury in bulbs and instrument displays

At the time of publication these restrictions apply from July 2003 and require re-engineering of products now on sale which continue to be sold after July 2003.

Although a cursory glance would suggest only limited relevance to the body structure many of the treatments are significant, e.g. hexavalent chrome being used in protective organic coatings and in paint pretreatments, solder and other joining processes, and alternatives are now actively being sought. Again each major company should have a manual which alerts engineers to potentially hazardous materials and which suggests alternatives.

Returning to normal in-house production procedures, precautions are be taken in handling materials at each stage of manufacture. During pressing detailed checks are carried out on the composition of coatings and press lubricants, and wearing of protective gloves is mandatory, the same applying to BIW assembly where adhesives are applied. In the use of epoxy adhesives robotic application is urged wherever possible and also in the pretreatment stages of paint application. With increasing use of galvanized steels, zinc fumes require extraction fans in effective use at each welding workstation especially where fusion welding is involved.

With current emphasis on environmental matters it is essential that each motor manufacturer maintains an awareness of developments in environmental legislation and it is now normal to devote significant resource to staffing a specialist function that both anticipates pending changes and communicates existing requirements to engineering and manufacturing areas. A typical aide mémoire issued to an engineering team provides a summary of a number of preferred practices – together with rationale – which should be adhered to by engineers.

8.6 BIW design for safety

As it can be considered as a further element of human protection, it is appropriate to include safety engineering within the same chapter as environmental control. Before focusing on the implications regarding body materials it is first necessary to briefly describe the nature of the tests so that the relevance of different material types can be assessed. This is summarized in a recent paper by Griffiths *et al.*[17] Historically, monitoring of crashworthiness tests has been carried out since the US New Car Assessment Program (NCAP) was introduced in 1978 and although following the same broad principles, differences do exist between requirements derived in the USA and those influenced by the European Experimental Vehicle Committee (EEVC) who administer EuroENCAP protocols. It is interesting to note that Australia recently aligned its ANCAP test procedures with those of EuroENCAP.

Basically the tests involve frontal and side-impact situations as illustrated in Chapter 2, Figs 2.11 and 2.12, and the conditions used for assessing compliance are summarized in Table 8.6.

As described by Griffiths *et al.*,[17] driver and dummy measurements are taken for the head/neck, chest, upper legs and lower legs in the offset crash test. Driver dummy measurements are taken for the head, chest, abdomen and pelvis in the impact test. Each of these body zones is assigned a score out of a maximum of four points, a poor performance rating zero and good rating four points. These relate to 'head-injury-criterion' (HIC) values of 1000 and 649 respectively using the alternative system. With the offset test the injury scores are subject to modifiers or penalty points, where, say, excessive rearward movement of the steering wheel might result in one penalty point applying to the head score (an unmodified score of four reducing to three). The modifiers include air-bag stability, steering column movement, 'A' pillar movement, structural integrity, obstructions in the knee impact area and brake pedal movement and therefore concerns regarding specific structural features can be highlighted for future attention. Both offset and side-impact scores are added to give a maximum of 32 points. The familiar star rating is based on overall score, for instance a rating of 22 points earning three stars. A typical EuroNCAP performance table is shown in Fig. 8.6. Table 8.4 summarizes details of European and US tests.

The ULSAB proving programme is an illustration of how compliance with NCAP procedures is achieved although CAE modelling was used to simulate the effect of displacement in physical barrier tests. Additional requirements to those already mentioned are also covered including rear moving barrier and roof crush. From a materials viewpoint it is useful to understand the deformation characteristics and collapse mode. In the 35 mph NCAP frontal crash the analysis shows good progressive crush of upper and lower structure with peak acceleration of 31 *g* and meets the AMS offset crash requirements. ULSAB also met standards for:

- 55 km/h 50 per cent AMS frontal offset
- 35 mph rear moving barrier
- 50 km/h European side impact
- Roof crush

Table 8.4 Crash test conditions and requirements[18]

Legal

	Europe	USA
Front Impact		**FMVSS108**
Speed	56 km/h	30 mph
Offset/angle	40% overlap	+/- 30°
Barrier	Deformable aluminium barrier face	Rigid face
Dummy	2 × hybrid III	2 × hybrid III
Criteria	Dummy injury levels only	Dummy injury levels only
Side impact		**FMVSS214**
Speed	50 km/h	33.5 mph
Offset/angle	90° to vehicle	63.5° to vehicle
Barrier	Deformable aluminium barrier face	Deformable aluminium barrier face
Dummy	Euro-SID	2 × US-SID (frt and rr struck side)
Criteria	Dummy injury levels only	Dummy injury levels only

Consumer testing

	Europe	USA
Front impact	**Euro-NCAP**	**US-NCAP**
Speed	64 km/h	35 mph
Offset/angle	40% overlap	+/- 30°
Barrier	Deformable aluminium barrier face	Rigid face
Dummy	2 × hybrid III	2 × hybrid III
	18 month child	
	3 year child	
Criteria	Dummy injury levels	Dummy injury levels only
	A-post intrusion	
	Brake pedal intrusion	
	Steering wheel displacement	
		IIHS
Speed		64 km/h
Offset/angle		40% overlap
Barrier		Deformable aluminium barrier face
Dummy		2 × Hybrid III
Criterial		Dummy injury levels
		A-post intrusion
		Brake pedal intrusion
		Steering wheel displacement
Side impact	**Euro-NCAP**	**SINCAP**
Speed	50 km/h	38.5 mph
Offset/angle	90° to vehicle	63.5° to vehicle
Barrier	Deformable aluminum barrier face	Deformable aluminium barrier face
Dummy	EuroSID (frt struck side)	2 × US-SID
	18 month child	(frt and rr struck side)
	3 year child	
Criteria	Dummy injury levels only	Dummy injury levels only

8.6.1   Influence of materials

It is generally accepted that energy absorption is a key requirement for any material to be used in impact resistant applications, typically front longitudinal members and side intrusion rails. Two factors affect performance, geometry and tensile strength, and

Latest results – Family cars

Model	Year	Front and side impact rating	Pedestrian test rating
Large family cars			
BMW 3-series	2000/2001	★★★★☆	★☆☆☆
Citroën C5	2001	★★★★☆	★★☆☆
Ford Mondeo	2001	★★★★☆	★★☆☆
Hyundai Elantra	2001	★★★☆☆	★★☆☆
Peugeot 406	2001	★★★☆☆	★★☆☆
Skoda Octavia	2001	★★★★☆	★★☆☆
Vauxhall/Opel Vectra	2001	★★★☆☆	★★☆☆
Volvo S60	2001	★★★★☆	★★☆☆
Mini MPVs			
Honda Stream	2001	★★★★☆	★★★★☆
Small family cars			
Alfa Romeo 147	2001	★★★☆☆	★★☆☆
Peugeot 307	2001	★★★★☆	★★☆☆
Superminis			
Rover 25	2000/2001	★★★☆☆	★★☆☆

Fig. 8.6 Latest results: Family cars

arguments can be made on behalf of steel based on an impressive array of grades giving varying stress/strain characteristics, as below. Using the general principle that the area under the stress/strain curve is proportional to the work expended, the latest grades of dual-phase (DP), complex and multi-phase (TRIP) steels appear to have an additional advantage over other HSS grades due to their combination of enhanced ductility coupled with a high work-hardening coefficient as shown in Figs 8.7 and 8.8.

As explained in Chapter 2 the DP properties originate from their duplex structure comprising ferrite with a 15–20 per cent fraction of martensite, the ferritic matrix maintaining good levels of ductility at the relatively high strength level imparted by the martensite and associated work hardening. In the case of TRIP the initial phases comprise a mixture of ferrite, bainite and retained metastable austenite the latter undergoing a strain-induced transformation to martensite during forming, maintaining strain hardening and postponing the onset of necking. This can be described as a 'reservoir' of ductility due to associated strain hardening accounting for the favourable

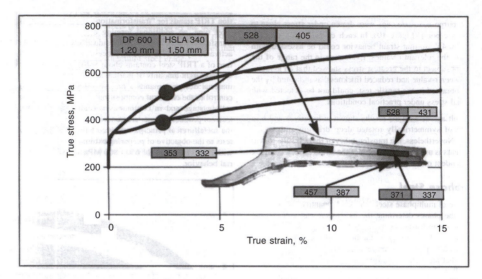

Fig. 8.7 Increase in yield strength by use of DP steel in comparison to HSLA[19]

enhanced ductility/high strength combination differentiating these from other HSS grades shown in Fig. 8.8.

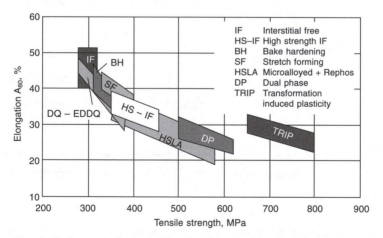

Fig. 8.8 Improved elongation levels associated with DP and TRIP steels[19]

The difference between DP and TRIP relates to the instantaneous *n* value where the behaviour of TRIP is more like that of austenitic stainless steel.[20] A further investigation[21] using 590 Mpa steels showed TRIP steels to have enhanced energy absorption in column collapse due to work hardenability and high '*n* value'.

Energy absorption has been demonstrated in axial (frontal) impact using box sections and side-impact situations using castellated test configurations. The weight reduction potential compared with other grades tested in both axial and impact bend tests was also shown.

The development and potential use of these materials appear increasingly attractive to the structural designer due to potential weight reduction and already significant utilization is evident. As described earlier by Ludke[22] up to 60 per cent of the body structure can benefit by using grades with yield strength values up to 300 Mpa (with future utilization approaching 80 per cent according to Tertel.[23]) Application was initially concentrated on front- and rear-impact resistant structural members with bake-hardening steels to improve the dent resistance of closure outer panels. Particular attention is now further being focused on side-impact resistance utilizing the DP and TRIP grades described above and the results of a SEAT evaluation using the grades in the applications depicted in Fig. 8.9 show a marked improvement in the behaviour of the DP/TRIP containing sideframe.[24] The use of other material forms is also helping to achieve improved safety standards. The BMW 3 series convertible is an instance where the strength of the roof was improved 70 per cent while maintaining the original weight, by the use of hydroformed tube parts.[25] The technologies used in the new generation of the GM Corsa vehicle include high strength steel, tailored blanks and ultra-high strength steel door beams as well as hydroformed parts.[26]

To the other grades of high strength steel can now be added materials with UTS values of 1000–1800 Mpa and while most other groups illustrated in Chapter 3 (Table 3.9) show sufficient elongation to enable them to collapse in a ductile mode in frontal- and rearward-impact situations these steels are more suited to sideways impact and common application is to intrusion rails in door assemblies.[27]

The steel composition and processing route can differ significantly for this type of steel and the first group of steels is typified by Thyssen Krupp Stahl who manufacture

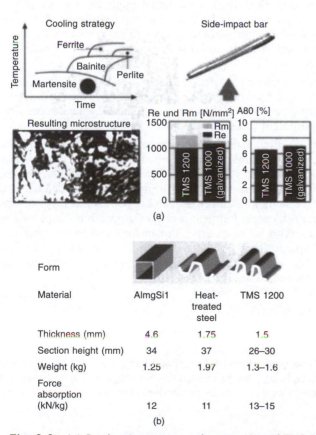

Form			
Material	AlmgSi1	Heat-treated steel	TMS 1200
Thickness (mm)	4.6	1.75	1.5
Section height (mm)	34	37	26–30
Weight (kg)	1.25	1.97	1.3–1.6
Force absorption (kN/kg)	12	11	13–15

(b)

Fig. 8.9 (a) Production route and properties of TMS 1200. (b) Comparison with other materials.[28] (Courtesy of Thyssen Krupp Stahl)

a steel with a predominantly martensitic structure (TMS 1200) using a specific cooling rate within the martensitic range as shown in Fig. 8.9(a), together with the microstructure. The claimed comparison with other materials used for side-impact sections is shown in Fig. 8.9(b).

The advantage over the other methods of steel manufacture at this strength level is that the blank material can be cold formed although tooling needs to be designed with precise allowances for springback to avoid the need for rework of the formed part. The alternative methods are described as follows and generally involve intermediate heat treatments which introduce higher costs.

Most of the heat treatable steels contain boron which through delayed transformation characteristics, together with a carefully balanced composition, give stability in the martensitic transformation. For Usinor Usibor 1500 steels a hot forming and quenching process is used, so shape during forming is accurately controlled and springback is not a major concern.[20] Achieving a durable coating on these steels can be a problem but Usinor claim a step forward with their Usibor AlSi product which utilizes the aluminized coating used for exhaust systems and which will withstand the austenitizing temperature and subsequent operations without scaling and therefore providing a receptive though rough surface for subsequent painting. A further variant is offered by Benteler, the German automotive supplier, with their BTR 165 grade.

A special mention should also be given here to materials such as Cellbond sandwich materials[32] of egg-box type profile, which are claimed might have a role in safety applications absorbing energy as each of the conical features collapses. This type of product may have an increasing application in the future and combines lightweight with potential safety in impact, but first the basic questions of panel fabrication and joining must be answered.

All the examples referred to above have featured steel parts, but aluminium is an undoubted contender in safety related body applications especially when 'mass specific energy absorption' is considered. This is illustrated in Fig. 8.10 where a comparison has been made between different materials for different parameters.[29]

The principles of adopting aluminium for body design and adapting for lower torsional stiffness have been discussed at length in Chapter 3, but regarding use for energy absorbing parts within a safety context, one of its merits is extrudability which can supply sections of very usable dimensions without the need for the welded seams that are required by steel sheet longitudinal and cross member box sections typically used for front- and rear-impact resistance. Extrusions allow flanges to be dispensed with and this means that parts with larger external dimensions can be accommodated within the same overall space. There is generally a need for thickening of sections compared with steel and by optimal section design a significant advantage can be gained over steel as demonstrated in Fig. 8.11.

8.6.2 F1 safety regulations

The requirements made of materials used in competition cars are illustrated by a further extract from Brian O'Rourke's authoritative article on Grand Prix F1 body structures[30] and although detailed allows the reader another insight into test procedures and how composite materials are able to meet today's demanding specification and contribute to the increasing levels of driver protection witnessed in recent years:

8.6.2.1 Testing

A protocol has been instituted within Williams Grand Prix Engineering Ltd regarding the criticality of parts. The failure of those which, it is considered, would jeopardize

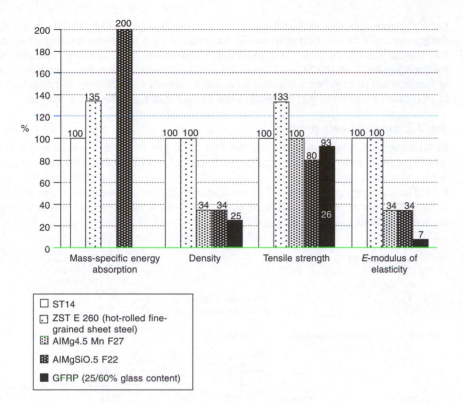

Fig. 8.10 Comparative properties of steel, aluminium alloys and glass reinforced plastic[29]

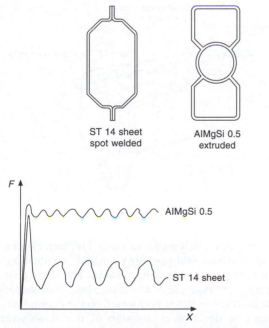

Fig. 8.11 Crush performance of relevant sections in steel and aluminium[29]

the structural integrity or handling of the car are deemed to be 'critical' and are designated as 'Class A' components. In accordance with good engineering practice, it is routine that all components or assemblies that are regarded as being 'critical' are, following release by the Stress Department, subjected to some sort of proof loading or function test prior to being released for use on a running car. Stiffness characterization is also carried out wherever deflection has been the design driver for the part.

8.6.2.2 Survival cell proving

In addition to the procedure above, the FIA have put in place a series of 12 tests which are intended to demonstrate that a minimum level of crashworthiness has been achieved within the design of the survival cell. They are illustrated in Figs 8.12 and 8.13. These are supervised by an FIA representative and must be carried out successfully before that model of car is allowed to enter Grand Prix events; in effect a 'type approval' system exists for F1 cars.

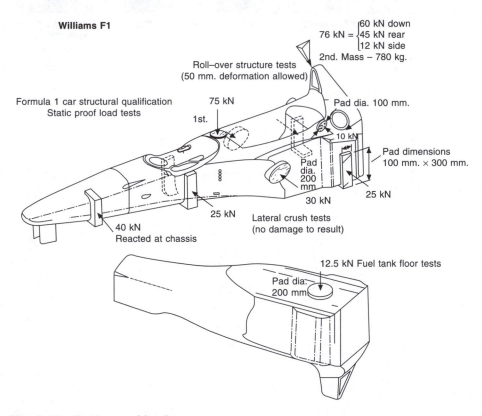

Fig. 8.12 Static proof loading

The tests may be divided into two types: static and dynamic. The former take the form of specified load levels with methods and positions for their application and reaction, whereas the latter are true impact demonstration cases. The static tests are further subdivided into those where every example of a structure built must be loaded and the remainder where a single destructive test is performed on the datum unit, i.e. the 'reference' structure that must be declared at the start of the manufacturing sequence and used for the complete set of tests. The others built must, when checked,

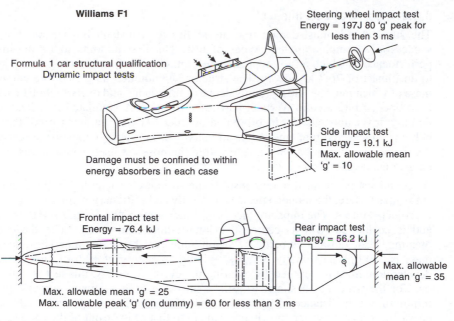

Williams F1

Formula 1 car structural qualification
Dynamic impact tests

Steering wheel impact test
Energy = 197J 80 'g' peak for
less then 3 ms

Damage must be confined to within
energy absorbers in each case

Side impact test
Energy = 19.1 kJ
Max. allowable mean
'g' = 10

Frontal impact test
Energy = 76.4 kJ

Rear impact test
Energy = 56.2 kJ

Max. allowable
mean 'g' = 35

Max. allowable mean 'g' = 25
Max. allowable peak 'g' (on dummy) = 60 for less than 3 ms

Fig. 8.13 Impact proof loading

show stiffness values that match, within a specified tolerance, those of the reference example.

8.6.2.3 Survival cell crush and penetration

The strength performance of the survival cell in terms of lateral crushing in checked by the application of 'squeezing' loads to one of its sides and reacted at the other. These tests must be carried out on all monocoques built and demonstrate that, during loading, deflections are contained within a specified level and that no damage results. There are four positions designated: the first is specified as being equidistant between the front wheel 'axle line' and the forward rollover protection structure; the second is aligned with the driver's harness lap-strap anchorage; the third is at a point on the cockpit edge related to the opening template; and the last is positioned to coincide with the centroid of the fuel cell compartment. The second position calls for 200 mm diameter circular load pads and 30 kN of proof load, the third 100 mm diameter and 10 kN, while the first and fourth require rectangular pads of dimension 100 × 300 mm for 25 kN loads. A fifth case is specified to check the fuel tank bay against penetration from below and calls for 12.5 kN to be applied on a 200 mm diameter pad, to be reacted through the engine mounting positions.

The three remaining static tests are performed once only on the reference survival cell. The structure must contain rollover protection features in front of and behind the driver. These must be proof loaded to demonstrate their capability and 50 mm of deflection is allowable in each case, meaning that some crushing of the composite structure is acceptable if it can be tolerated. The load in the first case is 75 kN applied in a vertical direction, whereas that for the second is a vector summation of three components (down, rear and side) totalling 76 kN. The third load is a simple lateral one of 40 kN applied to the nosebox which must not result in failure of its skins or connection to the monocoque front.

8.6.2.4 Survival cell impact

The survival cell must demonstrate its ability to withstand impact situations by successfully completing three types of test. The first pertains to frontal-impact performance in which a fully representative chassis and nosebox must be subjected to an impact of 76.4 kJ energy. The structure is mounted on a sled, the combined mass of which must be 780 kg (equivalent to a fully laden and overweight F1 car) and be propelled into a solid, vertical barrier at 14 ms^{-1}. A simulated fuel load and a dummy driver must also be incorporated in order to check the integrity of the seat-belt anchorages and fuel tank bulkhead junctions. This energy must be absorbed by the structure and contain the damage within the nosebox itself while not exceeding an average deceleration of 25 g.

A second test is intended to demonstrate the structure's capability in the event of a side impact. Here, the monocoque is held rigidly and stationary while a moving mass is projected into it. The impactor face is of dimensions 500 mm high × 450 mm wide and is positioned with its centre at a longitudinal station related to the cockpit opening, itself defined as being of a minimum size by regulation. The impacting mass is 780 kg, the velocity is 7 ms^{-1} giving an energy of 19.6 kJ, and the mean deceleration must not exceed 10 g during the event. In reality, this test is a difficult one for the structural designer to satisfy since the figures quoted correspond to a minimum crush distance of approximately 250 mm. The shapes allowed for the energy absorber, however, mostly correspond to that of the front of the car's sidepod intake which do not represent the ideal that might be had. The diffusion of the resulting loads into the chassis structure behind the energy absorber are, similarly, a challenge to design since the predominant load direction is normal to the monocoque side panel which must be supported laterally by some kind of substructure or internal beam.

Rear-impact protection is the third requirement of the technical regulations and is achieved by the provision of an energy absorber positioned behind the car's gear-casing assembly, usually doubling as a rear wing support structure. The test for this case requires that a gearbox, impact absorber, and rear wing assembly are fixed to a stationary barrier and a moving mass is projected into them. The mass of the impactor is 780 kg, the velocity is 12 ms^{-1} giving an energy of 56 kJ, and the allowable mean deceleration is 35 g. Again, for the optimum aerodynamic and mechanical layout of the car, the energy absorber component will usually be required to be minimal which presents challenges when maximizing its efficiency.

A final, fourth dynamic test case is that of the steering column which may involve composite materials in its construction. This calls for the steering wheel/column assembly to be subjected to a 200 J impact by a hemispherical object to simulate the driver's head striking it during a frontal accident. The failure criterion relates to an 80 g exceedance for more than 3 ms. This allows some scope in terms of design solutions as its location internal to the monocoque does not impose any geometric constraints resulting from aerodynamic considerations.

8.6.2.5 Imapct absorber design

Clearly, the impact conditions described above require considerable thought and effort when undertaking the task of designing the energy absorbers. The requirement to provide such components was first introduced for the season when the use of structural composites on F1 cars was still relatively new; the added task of making them deform and absorb energy during an impact was completely so. In the time period since then, however, a considerable amount of experience has been acquired in this field and has been advanced by the gradual extension of the regulations governing testing from the original, single, 39 kJ frontal impact ot the full spectrum of tests described above.

The question at the outset was – and today still is – how does one make use of 'brittle' composite materials in structures of this type to absorb the energy levels defined? It was found by constant experimentation that sandwich-stiffened carbon/ epoxy composite panels could, when subjected to end-wise loading, fail in compression in a manner which was stable and progressive, so providing a controlled retardation. This would allow us to achieve a useful force–time history characteristic during the impact event and so optimize the component for minimum mass or crush distance.

There are many factors influencing this crushing process which are necessary to understand in order to achieve an effective result: skin/core thickness ratio, overall panel size, material properties, failure modes and, above all, the geometry of the component. This design knowledge has been compiled over many years of work and different designs, during most of which the learning was empirical. In recent times and after considerable effort in research, at Williams' real success has been achieved with dynamic finite element modelling of the crushing process. The quality of the correlations found between simulation and reality has been such that it is now possible to explore many laminate variables before committing resources to the moulding of test pieces.

Since this article was written a side penetration requirement has been introduced for cars from 2001 onwards which involves producing panels which are representative of the survival cell laminate and subjecting them to penetration by a solid, loosely nose-shaped cone. For a load requirement of 150 kN the associated energy to penetration 100 mm is 6 kJ. Figure 8.13 shows suitably amended side-impact test 'energy' and 'g' values. The view of the original author is that this test has done significantly more to ensure uniformity of standards across the 'grid' and improve safety than previous requirements and is a really positive step for F1.

The growing application of FEA techniques was referred to in Chapter 4 on Formula 1 design but to prevent much time effort and material being wasted it would be useful to model the actual crash behaviour using FEA and this has been attempted with promising results by Gambling.[31] First, the behaviour of simple sandwich material was modelled (a sandwich tube of carbon fibre with aluminium honeycomb interlayer) and results then applied to an F1 nose structure. The initial correlation on deformation was good and then the same model applied to a full F1 structure, again with promising results. It may therefore be that with the validation of the DYNA3D models optimization of the F1 structure will soon be possible using FEA screen techniques in the workshop.

8.7   Learning points from Chapter 8

1. Reduction of the body mass and overall contribution to lower kerbweight have a major effect on the quantity of harmful emissions released to the atmosphere. This is especially relevant as the weight associated with safety and other product enhancing componentry increases progressively to meet market trends.

2. The trend to lighter structures is being encouraged by tax incentives such as that introduced in the UK whereby business cars are penalized according to CO_2 rating.

3. Selection of body materials increasingly involves the total life cycle to ensure that all environmental criteria are considered, e.g. energy requirements from production to disposal, and recyclability.

4. Metallic materials are relatively easily recycled and present only minor landfill problems. Polymeric materials will continue to present a more significant problem until rationalization of materials takes place favouring recyclable formulations and methods are developed for reuse.

5. Resolution of responsibility for vehicle disposal costs between manufacturing, retailing and governmental departments is essential if efficient ELV systems are to be introduced.

6. Recognized vehicle safety standard monitoring procedures are now evident in the form of ENCAP (New Car Assessment Programme) star-rating systems which summarize performance in frontal- and side-impact situations.

7. Materials such as dual-phase and TRIP steels are being increasingly used to improve crash performance by utilizing improved energy absorption characteristics resulting from higher strength and associated ductility levels. An extensive range of these multi-phase steels can now be considered as apparent from the ULSAB-AVC programme, and martensitic grades are available in strength levels up to 1200 Mpa for use in side-impact situations.

8. Different material forms are also being incorporated to improve the safety characteristics of new designs (a 70 per cent improvement in the strength of the BMW 3 series roof, also used to advantage in the new Corsa), utilizing hydroformed steel tube.

9. Extruded aluminium profiles can also be incorporated in designs to boost mass specific energy absorption, thus introducing the shape versatility of extrusions and also obviating the necessity for welded flanges.

10. Demonstration of the compliance of Formula 1 composite structures is required in accordance with detailed performance criteria and this is carried out on representative designs by testing at approved evaluation centres. FEA simulation techniques are increasingly being used in the prediction of F1 component behaviour.

References

1. Garrett, T.K., 'The Motor Vehicle', Butterworth Heinemann, Oxford, 2001.
2. Leggett, J., *Global Warming*, Oxford University Press, 1990.
3. Peugeot, R., Introductory talk, 'Steel and Automotive Body' Usinor Symposium, Cannes 14/15 June 2001, p. iv.
4. Anon, 'Faster than a Moving Tax Break', *Sunday Times*, London, 30.12.01.
5. Frankel, A., 'Step on the Gas', *Sunday Times*, London, 30.12.01.
6. Gaines, L., 'Lifecycle Analysis for Automobiles: Uses and Limitations', SAE Publication SP-1263, Feb. 1997.
7. 'Competition Between Steel and Aluminium for the Passenger Car', IISI, Brussels, 1994.
8. Peters, T., 'Environmental Awareness in Car Design', SAE Publication SP-1263, Feb. 1997.
9. Fox, J.W., 'Hypercars: A Market Oriented Approach to Meeting Life Cycle Environmental Goals', SAE Publication SP-1263, Feb. 1997.
10. Price, R.W., 'Measuring Consumer Response to Environmental Pricing in the Automotive Industry', SAE Publication SP-1263, Feb. 1997.
11. Gick, M., 'End of Life Vehicles – What Next', Automotive Composites Workshop Plenary Session, 2/3 Dec. 1998, Institute of Materials.
12. ACORD Fourth Annual Report, SMMT London, 2001.
13. 'The Disposal of End-of-Life Vehicles', Report by Knibb Gormezano and Partners, 30 May 2001.

14. Matthews, A.E. and Davies, G.M., 'Pre-coated Steels Development for the Automotive Industry', *Proc. Instn. Mech. Engrs,* Vol. 211, Part D, 1997.
15. 'BMW Group Recycling – Trendsetting and Convincing', BMW Group publication, Munich, 2001.
16. Broge, J.L., 'Several Steps Toward Recovery', *aei Magazine*, May 2001.
17. Griffiths, M. *et al.*, 'Consumer Crash Tests: The Elusive Best Practice', Worldwide Harmonization of Crash Test Programs Symposium, Cologne, 2/3 Dec. 1999.
18. Green, J., Private communication, 2001.
19. Hartmann, J. *et al.,* 'High Strength Steel Sheet for Automotive Design', IBEC '97 Proceedings, Stuttgart, Germany, 30 Sept.–2 Oct. 1997, p. 65.
20. Continieaux, P., 'Steels with UTS levels > 800 Mpa', 'Steel and Automotive Body' Usinor Symposium, Cannes 14/15 June 2001.
21. Mizui, N. *et al.*, 'Fundamental Study on Improvement in Frontal Crashworthiness by Application of High-strength Sheet Steels', SAE Publication SP-1259, Feb. 1997, pp. 39–44.
22. Ludke, B., 'Functional Design of a Lightweight Body-in-White in Steel Taking the New 5-series as an Example', Usinor Symposium Acier et Carosserie Automobile, Cannes 14/15 Sept. 1995.
23. Tertel, A., 'Weight Reduction – Solution Implemented in the New BMW 3-series', Usinor Automotive Symposium, 1999.
24. Bekemeier, F., 'Use of EHYS in New Vehicles', 'Steel and Automotive Body' Usinor Symposium, Cannes 14/15 June 2001.
25. Tertel, A., 'Use of Hydroformed Steel Parts in the Body of the New BMW 3-series Coupe', 'Steel and Automotive Body' Usinor Symposium, Cannes 14/15 June 2001.
26. Mengel, C., 'Body Development of the New Opel Corsa', 'Steel and Automotive Body' Usinor Symposium, Cannes 14/15 June 2001.
27. Rostek, W., 'Applications of Ultra High Strength Steels for Chassis and Impact Components', 'Steel and Automotive Body' Usinor Symposium, Cannes 14/15 June 2001.
28. 'Side Impact Bars for Cars from Ultrastrong Hot Strip', *TKS Compact* magazine, Jan. 1998, Thyssen Krupp Stahl AG.
29. Ostermann, H. *et al. Aluminium Materials Technology for Automobile Construction*, Mechanical Engineering Publication Limited, 1993.
30. O'Rourke, B., 'Formula 1 Applications of Composite Materials', *Comprehensive Composite Materials*, Elsevier Press, Oxford, 1999.
31. Gambling, M., 'Simulation Techniques for Impacted Composite Materials with Reference to Formula 1 Car Applications', Automotive Composites Workshop, Brands Hatch 2/3 Dec. 1998.
32. Anon, 'Crunch Time', *Materials World*, Vol. 9, No. 8, Aug. 2001, pp. 16 and 17.

9 Future trends in automotive body materials

Objective: The future utilization of materials is considered as evident from trends, opportunities arising from developments in materials technology, and also the possible effects of legislation. As well as proposing scenarios likely under normal conditions, possible options arising from a sudden change in circumstances are considered, e.g. by a significant reduction in oil resources, or the need to respond more urgently to environmental pressures.

Content: General prospects of change within a highly conservative industry are considered first – primary and secondary influences are evaluated – the effects of legislation resulting in a need for reduced fuel consumption/alternative fuels and greater recycling efficiency are reviewed – improvements affecting materials choice resulting from improved manufacturability are identified – options are then proposed which could arise within normal/shortened timescales for high and low volume production situations.

9.1 Introduction

At the outset of any discussion on future trends in automotive body materials it must be said that the volume car industry is very conservative, and exceptional circumstances will be needed to bring about any significant change. Economics are a prime consideration in mass production and this has been covered in detail at the end of Chapter 2, where it was demonstrated that steel was still the main choice under current conditions and significant changes in manufacturing strategy or raw material cost would be needed to effect any real shift from this preference. Regarding the more esoteric materials, circumstances would have to change dramatically if the additional weight reduction they offer is to be realized. Although presenting the prospect of a natural renewable source of material, realistically it is thought that any trend to utilize natural materials or fibres will be strictly limited. Likewise despite the progress made with metal matrix composites and forgings with chassis and power train parts, it is unlikely these will complement sheet, castings, extrusions and tube within the body structure, and unless an economic breakthrough is made, even carbon fibre composites will struggle to find a wider application. However, niche and premium car sectors may respond differently to the factors which could stimulate any change and should be viewed separately, and therefore it may be useful to review likely changes in both volume and niche car categories, with regard to longer-term expectations and more urgent implementation which may be brought about unexpectedly by external factors. Given normal circumstances the primary areas of influence bringing about change are likely to be, first, fuel availability and type, which as we have seen in Chapter 8 is strongly related to emissions control, and other environmental pressures such as recycling plus legislation surrounding the ELV situation. Other important but secondary influences include more stringent safety engineering requirements that might arise, and also any developments that are made in the field of automation and manufacturing technology that may favourably influence the choice of certain materials and contribute to significantly lower costs. As mentioned previously it is these factors that have favoured the selection of steel, its versatility ideally suited to robotic processing, with improved HSS grades and coated derivatives that can be easily accommodated within the range of processing equipment. Recycling presents few problems. Reference is also made to developments (or future development prospects) which may further prolong its implementation.

The timescale is critical and any sudden threat of fuel shortage could hasten the need for lightweight structures. However, it is possible that given long enough, the situation could favour the status quo. Time is required to resolve political issues such as safety vs weight reduction ('flimsiness') in the USA, and European issues concerning provision of alternative fuel pump infrastructures (e.g. for hydrogen), funding of ELV, etc. Indeed it appears from the most recent Earth Summit conference in Johannesburg that the USA is reticent with regard to environmental commitments, as too are some other industrial nations who see sustainable economic growth as a priority. Therefore, combined with the slow inertia accompanying change within the industry, it is probable that unless unforeseen circumstances do arise, changes will be gradual allowing the development of alternative fuel systems in a more efficient form. Given this intervening period and time for common sense to prevail it is also probable that 'sensible' solutions will be found to outstanding recycling, plastics rationalization, and processing issues. This would then allow the continued use of steel, but now perhaps complemented by increased utilization of plastics, and aluminium where appropriate. Using mixed materials in this way would allow the manufacturing advantages and inherent safety associated with steel structures to be combined with the shape versatility of plastics (e.g. 'friendly' front ends), while using aluminium to shed weight (e.g. closures). It is by hybridizing the structure in this way and exploiting synergies available from such systems that the importance and versatility of materials technology will be increasingly recognized in the vehicle bodies of the future.

9.2 Factors influencing material change in the future — trends and requirements

External influences are considered first in the following review, reflecting the high priority which the industry gives any topic of public concern. These include environmental issues and the increasing effects these may have on materials selection through emissions control and recycling legislation, but questioning the need for drastic change if alternative fuel cell technology is imminent. Profit generation is obviously the lifeblood of the industry and the effects of globalization and economies of scale on design and the consequent ability to absorb change quickly are next considered, together with advances in computer-aided engineering procedures which could shorten model programme development times significantly. Reference is made to the Audi A2 which demonstrates the mass production capability of aluminium, and the possible future increasing use of carbon fibre composite structures which allow strengthening and stiffening exactly where required with little material wastage. The development and demonstration of compatible manufacturing systems to accommodate newer materials within normal budgetary constraints would help the introduction of lighter weight alternatives to steel and any product improvements easing processing of materials will favour their selection. Examples of such improvements and future areas of potential benefit which would improve processing and implementation are identified. The need to fully evaluate new processing techniques which are essential to extract best use of materials cannot be emphasized enough but instances could be quoted where current model programme introduction has been seriously delayed by under-rehearsed manufacturing procedures. Already significant changes in material utilization are evident, especially in niche car developments, and the ways in which all these influences may crystallize in future, and for differing conditions, are suggested in a final table showing model precedents where appropriate.

9.2.1 Influence of environmental controls

As apparent from Chapter 8 the automotive industry features prominently when environmental issues are discussed and both emissions controls and recycling targets have been introduced at national levels. Efforts made in this respect by individual companies, automotive organizations and governmental departments have been publicized and it would be expected that these should influence materials choice in time as lightweight and reusable alternatives are adopted. However, organizations representing existing fuel, material and other interests will tend to oppose any significant change, and with the protracted effects of political inertia accompanying such changes implementation will tend to be slow, favouring the status quo and in the next section the *real* prospects for these issues contributing to change are examined.

9.2.1.1 Emissions control and fuel systems

Despite fuel consumption being a main concern since the 1970s when legislation in the USA dictated that specific corporate mpg levels were reached by companies distributing vehicles in that country, and more globally thereafter as pressure has mounted for emissions control, very little sign of commitment on behalf of the consumer is evident over 30 years. 'Gas guzzlers' still abound and the popularity of 4×4 and other large vehicles appears to be growing. Car companies are making real efforts to meet legislative controls by producing models with emissions levels reduced below 100 g/km, but is there a real commitment from the public to drive such vehicles without stronger controls being imposed? Comfort and practicality appear to be of similar or higher priority and it is not until more drastic legislation or tax regulations are introduced that any noticeable effect may be evident. However, the commitment from the manufacturers is clear – with evidence that all major companies in the USA are demonstrating a capability of 80 mpg with development vehicles, Japan is already producing cars with hybrid electric/petrol power units and in Europe companies such as BMW have proven they can adapt to hydrogen for regular designs such as the 7 series.

A key issue will be the type of fuel system to be adopted, as the more cumbersome these vehicles are then the more weight compensation has to be contributed by lightening the supporting structure. It is not necessary to consider future propulsion systems in detail here but within the UK and Europe there appear to be two scenarios developing,[1] one of which foresees the phased introduction of increasingly more hybrid systems until 2020, and the other where hydrogen-based systems might be more predominant by 2050. Progress achieved already with ten European models already showing CO_2 levels below 100 g/km suggests that current targets are within the scope of existing engine technology. Hybrid technology, whereby current systems are augmented by electric power, should then ensure further reductions if necessary, until hydrogen and fuel cell type power units will see CO_2 emission levels drop to almost zero anyway.

Therefore a smooth transition between these technologies appears possible within the timescales indicated and this should be accompanied by complementary changes in materials utilization. As we shall see later solutions appear available for even the heaviest of the newer electric driven systems, but development of manufacturing systems to accommodate these will then become another key requirement.

9.2.1.2 Recycling and ELV legislation

From statistics on the recovery of materials produced by the ACORD group, the recycling of metallic materials appears to present no major concerns for the future.

The obstacles that impede progress in the utilization/recycling of polymeric materials have been discussed in Chapter 8 but if the obvious advantages of plastics are to be exploited more fully a number of issues must be addressed more urgently.

These include rationalization and here the example of the formulation and reformulation of national and US/European standards and the emergence of ISO worldwide specifications with regard to metallic materials provides an example and has undoubtedly encouraged convergence of similar grades and alloys. Within Europe, the slight differences which were in evidence with early rephosphorized variants have now been standardized within EN 10292 and the same is happening as standards for newer steels such as the multi-phase formulations. Most 'in-house' automotive specifications used in body production are linked to these standards, resulting in advantages in consistency of product, processing and economics and has allowed easier discussion on common environmental issues across the industry. Alignment of supplier products with automotive specifications has been needed in the case of some aluminium sheet products but this has helped processing uniformity and recycling, which may have been a problem in Europe. For example, Al–Cu alloys, which fitted comfortably within some suppliers' product portfolio, fulfilled functional requirements of skin panels but did not meet recycling requirements, and have now been replaced by Al–MgSi even if alternative sources of supply had to be found. This experience may be the key to recycling of plastics and already it has been identified[2] that over 20 polymer types are used in automotives but six of these account for 70 per cent of the weight of plastics, namely, PP, PUR, ABS, PE, PVC and PA, and that concentrating on these would simplify all aspects of recovery including identification, dismantling and segregation, and reuse. The specification of preferred, recyclable materials such as polypropylene will then help to elevate recovery targets to nearer 90 per cent (a declared aim of the BPF for 2015) from the existing level of 75 per cent.

More immediately the accessibility of material specifications to dismantlers is already being expedited electronically by car manufacturers, and allied to this are the development of identification techniques such as the triboelectric pen and adapted FT infrared spectrophotometry developed at Southampton University with Ford,[3] which again will simplify the dismantling process. Examples of techniques used for reuse of polymers such as skin flotation techniques to enable separation of ASV shredder residues and resin development to unzip the molecular structure into its constituent monomers for rebuilding new, e.g. Nylon 6 polymer, have been referred to in Chapter 8 and such research programmes must continue. However, the motivation to move all these initiatives forward is funding and until it is decided where the responsibility for disposing of vehicles lies, and availability of resources for segregation and recycling identified, progress will be limited. Unless resolved quickly the application of polymers for body panels will not exceed growth rates observed within the last 10 years.

9.2.2 Future effects of design and engineering trends on material selection

To maintain profitability within larger organizations it is essential that economies of scale must proceed on a global scale and this has witnessed the international marketing of similar if not identical models. The commonality is now extending to the use of the same basic platform or floor pan which can be modified to accommodate many of the variant types of model made by a particular company. This is not necessarily restricted to a particular model size and may encompass sedan, sports car or utility formats. This is illustrated by reference to the recently completed ULSAB-AVC programme[15] where common parts were used for a European type mid-size design (C-class) and US counterpart (PNGV-class) as shown in Fig. 9.1.

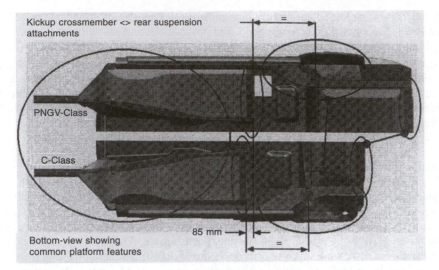

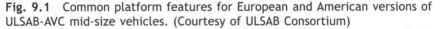

Kickup crossmember <> rear suspension
attachments

PNGV-Class

C-Class

85 mm

Bottom-view showing
common platform features

Fig. 9.1 Common platform features for European and American versions of
ULSAB-AVC mid-size vehicles. (Courtesy of ULSAB Consortium)

This trend will continue and many of the 'world players' with manufacturing plants
in many countries would wish to retain the flexibility and design technology associated
with tried and tested materials, for which local supply and servicing infrastructures
exist and this favours steel. The challenge for aluminium and suppliers of other
materials is to match the 'universality' of these easily adapted structures for the
future. Parts of a spaceframe can be easily shipped in KD form to foreign countries
but can these be modified, adapted and serviced as easily locally – especially when
future variants require cut down or enlarged versions? Therefore another aspect of
'global inertia' will further stabilize the current trend to progressively develop and
extend the use of steel. Alternative systems, although proven for mass production by
some companies whose trade mark is in innovative development, are not yet robust
enough to deliver the wider demands of the industry. Unless economics change
dramatically therefore or stronger environmental forces prevail, steel will prevail for
numbers production for a longer time than might be suggested by the many debates,
initiatives and forecasts that have taken place over the last 30 years.

Exceptions exist as where sales of an international best seller are contingent on local
conditions dictating a high local labour content and an example is where the bodywork
of the original Mini was produced in GRP plastic for South American sales. As
stated, the rapid development and deployment of CAD resources will result in ever-
shortening lead times in the future and allow the predictive behaviour of most aspects
of design to be simulated without recourse to the costly iterative engineering programmes
still carried out to assess many situations. This particularly applies to impact simulation
which is already used widely, even for Formula 1 as described in Chapter 4 where
very close predictions are already being made for composite structures. Fast deformation
programmes available for modelling behaviour of crash situations for road cars are
now being complemented by systems predicting the forming behaviour of different
metal sheet grades during press forming. With increasing computer capacity there is
no reason why design programmes should not extend to include material behaviour
in body structures allowing substitutional exercises in the selection of alternative
metallic/non-metallic materials and forms. This would clearly indicate the potential
advantage of possible substitutions and allow the designer another option (i.e. choice

of a different material) when local strengthening is considered – exactly where required – as in TWB or F1 carbon fibre construction. This does highlight one problem and that is the consistency of data banks regarding presentation of material properties for numerical systems. Many sources of information on physical and mechanical properties exist but units for metallic materials can differ according to form (e.g. elongation values for sheet or castings) and comparison with the many forms of polymeric materials is again difficult. This highlights the role of central organizations such as the Institute of Materials based in London, in promoting the production rationalization and distribution of such data banks for unless the input data is uniform and reliable even the best of predictive systems will prove inaccurate. The provision of this information technology is one of the key enabling factors dictating the extent and pace of change for the future.

Longer term it can be expected that CAD will play an even more effective role in shortening model timescales from concept to production although it is unlikely that techniques such as those used for rapid prototyping whereby laser and resin systems can produce 3-D shapes will ever be used for production parts.

More tangible improvements in design will include increased use of laser welding as this is likely to become more widespread as transmission systems and panel accuracy improve as described above. This will provide stiffer joints replacing spot welded arrays with linear seams, in particular by application to peripheral roof joints and also around door frames to augment hem flanges. Not only does this stiffen and allow a degree of downgauging but the continuity of the weld dispenses with the need for an adhesive. As accuracy improves and line following techniques are optimized, the weld can also be placed at the end of the outer panel fold to displace the edge sealer.

Again it will be the most versatile of materials that will be able to respond to the use of such systems and feature in future models.

9.2.3 Advances in manufacturing technology

Enhancement of material features benefiting production efficiency and processing costs will also help accelerate the acceptance of new materials and before some of the more radical changes (e.g. the adoption of extrusions, castings, hydroformed tube and sandwich or honeycomb forms) are considered, some of the more fundamental development trends and requirements which could influence the shorter range choice of materials are identified.

9.2.3.1 Forming

The 'process chain' involved in the production of automotive bodies comprises a series of forming, assembly and finishing operations all of which have become increasingly automated. The trend in press forming, still by far the predominant method of shaping body parts, is to use increasingly large tri-axis or progression presses with internal transfer mechanisms to maximize productivity. Consequently, any disruption due to material performance can cause massive numbers of panel rejects before, during and after detection and therefore any change must be very carefully planned, rehearsed and monitored before even partial runs are attempted. Without demonstration of a successful condition, substitution of materials cannot be recommended, and suppliers must be fully enrolled to ensure the agreed performance can be maintained in production. This applies to steel grades where modified coatings and alternative grades of high strength steel represent the most recent type of changes implemented. Despite the accumulated experience with sheet steel, even these require extended proving times as coating pick-up and lower ductility levels have to be accommodated. The investment level in these facilities is high and can only be justified if high throughput levels are maintained. In these circumstances the introduction of a different material such as aluminium with inherently lower formability, and for which much more careful housekeeping and interpress cleaning procedures are necessary, causes problems. It is difficult to run both steel and aluminium on the same lines, as wash oils, cleaning procedures and other disciplines are different, and as the trend to hybrid or mixed body conditions continues dedicated tandem lines will be essential for each material. The alternative could be extensive and very expensive rework!

As a result intensive development work is already taking place to ensure as much surface preparation takes place before the material leaves the supplier, but one fundamental requirement that will ease mass production of aluminium pressed parts is the availability of coil together with the proven ability to handle and blank it with minimal damage, as much of the material currently used is in sheet form.

To preserve the surface of aluminium during transit (particularly fretting damage which can arise through faying surfaces), pressing and handling, and to optimize the paint finish, prefinished aluminium strip will almost certainly be required in increasing quantities. For maximum consistency this should feature the EDT finish and a wax that enhances lubricity but which resists sticking during blank destacking prior to being fed into the press, and is compatible with later applied adhesive systems. For painting the environmentally preferred Ti–Zr (or similar) formulation could also be applied as a pretreatment at the coil stage before leaving the supplier, and the feasibility of supply to such a specification has already been demonstrated. As mentioned a little later for steel this precoating technology can be applied up to the prepaint stage if necessary and this again must be an area of increasing assessment activity, and could yield massive cost savings resulting from the deletion of in-house paint facility costs. This highlights a further increasing trend in the future – the transfer of much of the secondary processing of material back to the supplier. This enables any problem solving to be undertaken by the initial experts, expedites assembly as suboperations are handled off-line and removes some environmental issues, e.g. fumes, effluent disposal, etc., from the automotive plant.

9.2.3.2 Assembly and finishing

The same need for radical change is required in assembly facilities to accommodate new materials. Although robotic manipulation and processing allows some flexibility regarding model variants and welding settings, and can adapt to the range of steels mentioned above, other potential alternatives again present a need for significant investment. As explained in Chapter 6 aluminium requires increased transformer capacity which hampers miniaturization and robotic manipulation, and much more careful handling procedures.

Laser welding systems should advance significantly during the next decade to allow far more controlled joining operations to be used in volume production, with improved accuracy, reduced distortion and less oxidation/coating removal in the case of zinc coated steels. CO_2 lasers have proved cumbersome in the past, requiring mirror systems to carry the beam to the location on the workpiece and overall reliability in production situations has been questioned. Now that YAG and other systems are being developed which can be transmitted by fibre optics, these carry the prospect of much more practical and usable application. However, these need touching contact and more consistency from the press shop is required in the form of accurate flange shape and flatness to enable their effective use without relying on flattening wheels, etc. which limit access to certain conditions. Thus another requirement is that materials can respond to laser power sources and although aluminium suffers from its reflectivity this is being addressed, and this is a mode of joining which can also be applied to plastics. As well as being the primary mode of assembling tailor welded blanks, laser welding could also be used to join hydroformed sections and again accuracy is required to ensure that any 'stack-up' of accumulated gaps from misfitting parts can be taken up.

Painting is more accommodating and modern processes are designed for mixed metal steel/zinc coating/aluminium combinations, although increasing hybridization and the inclusion of plastic panels in the body shell may cause fitment problems due to differential expansion characteristics, requiring loosening at BIW before final tightening during the trim and finishing stage. The development of in-mould painting to required body colour and attachment of plastic parts at a later stage in the process would help the introduction of plastics, where difficulties in initial applications are proving a deterrent to the specification for body parts.

Therefore considerable commitment will be needed for the significant change in materials in the future. The experience of Audi in changing the design protocol and material format in mass producing the all-aluminium A2 illustrates how extrusions and castings can be used to overcome some of the previous problems but does also show the size of investment needed to re-equip for these new forms. It is far more difficult to imagine the type of process required to mass produce ultra-lightweight carbon composite, honeycomb and sandwich construction modes that might be required by electrically driven cars, but it is essential for credibility of the widespread use of these materials that this area of manufacturing technology must be researched, in-depth, now.

9.2.4 Improvements in materials specification – trends and requirements

Before forecasting the changes that might occur with regard to the substitution by other materials, enhancement of the characteristics and versatility of existing materials could lead to their longer term retention, particularly if use of existing production facilities can be prolonged. Again these shorter term developments and requirements are identified before moving to more radical possibilities.

9.2.4.1 Pre-coated sheet

The most immediate change imminent with regard to coated steel concerns the advances in duplex coatings, which consist of a zinc coating usually electrogalvanized with a superimposed organic coating. 'Bonazinc' is typical of this type of coating and comprises a zinc layer 5.0 microns thick with an overlayer of weldable primer. This type of coating is starting to be implemented within European bodyshells such as the Daimler-Chrysler 'A-Class' currently for door construction[4] with claimed advantages regarding corrosion and facility savings. The duplex coating provides improved corrosion resistance and allows deletion of the cavity wax operations and savings achieved through dispensing with robotics more than offset the initial cost premium of the material. Extending the concept to the replacement of the 'in-house' painting process applied within the car plant by the use of strip with the paint partially or wholly applied at the steel mill holds an exciting prospect for vehicle manufacturers and this is now well within the bounds of possibility. Already such materials are utilized for the production of domestic appliances. Replacement of the priming process alone, dispensing with the need for an electropriming facility, might save investment costs approaching £20M per model while also reducing some of the environmental concerns which normally surround an in-house paint facility. Trial pressings made from development material have already demonstrated that given careful handling in the press shop the integrity of the paint film can be maintained and an outer panel final paint finish obtained with this material as shown in Fig. 9.2.

For closures and spare parts, where clinched joints and mechanical fastening can be used to replace welds and hide cut-edges, this technology has already been proven feasible. The main problem still lies in joining the material and it is essential that for widespread adoption a weldable formulation must be derived, and this is the current challenge before moving to develop complete fully finished paint systems.

9.2.4.2 Zinc coated steel – PVD coatings

The next change with respect to zinc coating steel technology relates to the perennial problem of pick-up of zinc particles/build-up of zinc on press tools, and accelerated electrode tip wear due to alloying of molten zinc with the copper-rich spot welding electrodes. These aspects must be addressed if efficiency of all links of the process chain are to be improved. In a recent BRITE-EURAM collaborative project[6] it has been demonstrated that using physical vapour deposition techniques coatings as low as 4–5 microns can be deposited which give advantages with respect to both press performance and electrode tip life. Although presenting a challenge in terms of the production of strip wide enough for automotive production this has already been shown as a viable alternative method of zinc deposition in Japan[7] and the economics approach those of the electrogalvanizing process. The advantage of the process as proven in the joint European project was that alloyed layers of zinc plus elements such as Ti, Cr and manganese could be deposited which proved to show enhanced corrosion resistance compared with standard automotive quality zinc coated steel. The direction of this work must now be redirected to optimize preferred alloying elements, the format of sandwich layers, and regarding the processing parameters such as the nature of the vapour itself. Initial indications show that this unique method of alloying could eventually lead to a product which answers all the questions posed – cost, coating integrity, corrosion resistance and processability.

Fig. 9.2 Rover door outer panels and wing and using formed using preprimed steel sheet

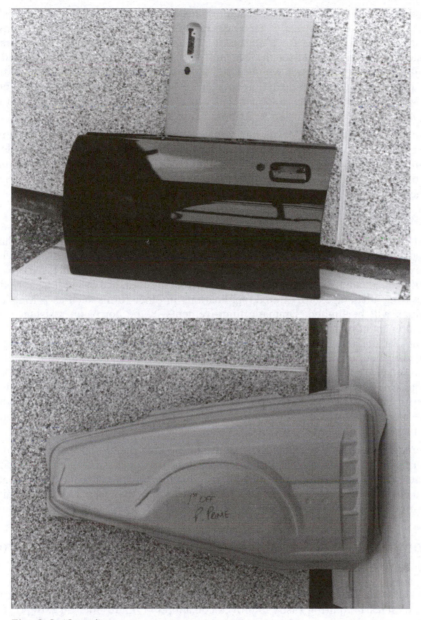

Fig. 9.2 *(Contd)*

9.2.4.3 High strength steel range

The modern steelmaking procedures already described in Chapter 3, including the capability to heat treat strip in-line with accurate cooling rates, allow the production of an increasingly wide range of dual-phase, TRIP, multi-phase and martensitic steels. In future this expanded variety of grades should allow more efficient use of material designs and this is illustrated by reference to the ULSAB-AVC programme[15] which used the grades shown in Fig. 9.3 and which participating steel suppliers agreed would be available to use in models meeting 2004 regulations.

Through synergistic use of these 'enabling advanced high strength steel grades' it was possible to demonstrate 5-star crash ratings and economy figures of 84 mpg (diesel highway driving) for a European 'C class' (VW Golf type) vehicle and considerable enhancement for the US mid-size 'PNGV-class' model considered in parallel. The body structures weighed 202 kg and 218 kg respectively, representing a 17 per cent reduction compared to an average benchmark. It is interesting to note that weight reduction of up to 42 per cent was claimed for outer panels through the use of hydroformed steel down to 0.6 mm using sheet hydroforming.

It would be expected that development of low carbon steel sheet would continue, using careful control of composition and heat treatment to provide higher levels of ductility commensurate with strength. It is unlikely though that the engineers' ultimate aim of increased elastic modulus, while retaining other key properties, will be realized in the foreseeable future.

9.2.4.4 Potential for novel materials and forms

Most realistic materials developments which could be adopted for future use have already been covered in the main text and extend to castings, extrusions, sandwich, honeycomb and other forms of material. As already explained, unless there are significant advantages in strength, density, corrosion resistance or modulus, or other processing advantages are offered, applications in other forms such as MMCs and forgings will be limited. Dramatic improvements in the corrosion resistance and development of higher strength magnesium alloys have made these more attractive in recent years for aerospace use (e.g. WE43 gearbox castings for the McDonnell Douglas MD500 helicopter) and forgings enable use at high temperature, although little use is foreseen for autobody parts due to economics. Some use of the RZ 5 alloy has been adopted for Formula 1 racing cars[8] for gearbox covers and MSR/EQ21 alloys due to superior ambient temperature properties or high operating temperatures, but while these meet arduous racing car requirements this is currently of limited relevance to the majority of mass produced cars.

Reference has already been made to the use of the high ductility levels exhibited by superplastic materials[9] and the limited use made of these materials for panels such as bonnet skins. The feasibility for full body manufacture has already been demonstrated and a cost effective case utilizing cheap concrete tooling may be made where limited investment might be involved for export markets.

9.3 Combined effect of above factors on materials utilization within 'expected' and 'accelerated' timescales

Unless there is some form of catastrophic change demanding an alternative to current fuels or at least requiring vastly improved consumption figures then an extrapolation of change rates evident over the last 30 years can be expected, i.e. steady change. As for the emergence of galvanized steel in the 1980s through competitive pressures triggered by awareness of Audi demonstrating the feasibility of 100 per cent utilization, the next stage might see a progressive utilization of aluminium substructures by premier car divisions of larger car corporations (to combat thirsty larger engines) clad with aluminium closures as dent resistance, hemming performance and surface consistency, aided by pretreated strip, improves. Given further time it is likely that polymer panels will be more prevalent once rationalization of different types is achieved, a Class 'A' finish (perhaps self-coloured or in-mould coated) and recycling efficiency reaches an acceptable level.

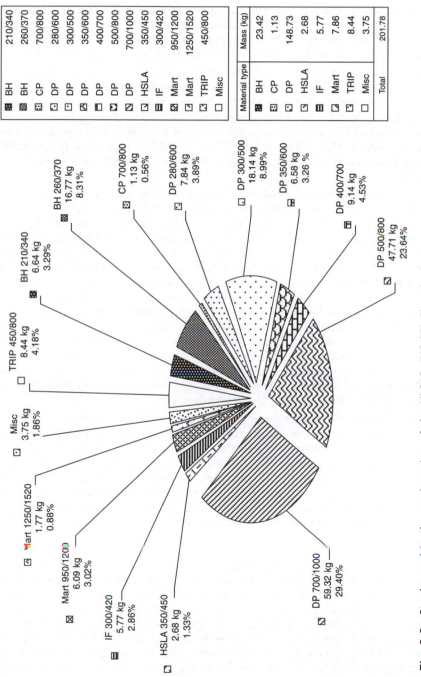

Fig. 9.3 Steels used in the engineering of the ULSAB-AVC 'C' Class body structure

Questions still remain unanswered as to the benefits of aluminium. There is, as yet, no convincing proof that the use of aluminium will substantially change the rate at which legislative emissions targets are met or contribute to significantly enhanced fuel consumption figures for volume cars. Certainly the power to weight ratio of the NSX and other high performance cars is boosted by the use of lighter materials but apart from special variants of the model can the A2 boast significantly better fuel economy than some of its rivals? Despite the interesting and technologically successful design of this car the overall rationale for such a car at the present time has been questioned by experts in the field[10] who initially considered the A2 to be significantly overpriced and refer to 'the mind boggling £17 000 for an A2 – probably £10 000 more than it's worth.' No doubt others in this field take an alternative view.

For volume cars it is more likely that steel-based structures will still predominate with an increased trend to precoated zinc or prepainted coatings once joining technology has advanced sufficiently to exploit the advantages of this material. Again once the basic hurdles to recycling have been removed the widespread use of polymer panels could materialize, and continuing production of the GM Saturn derivatives indicate this is possible. Hence the summary chart of Table 9.1 presents an impression of likely options, dependent on operational conditions. Table 9.2 summarizes the overall trends.

This is meant to provoke/stimulate discussion rather than be a definitive forecast, but nevertheless may contain some realistic ideas for the future. Examples are given of where specific technologies have been applied on production models. Again these are not necessarily the most up to date uses, but show a reasonably topical application. For volume cars it is more likely (Option 1) that steel-based structures, which fit comfortably within existing/easily modified facilities, will still predominate with an increased trend to precoated zinc or prepainted coatings once joining technology has advanced sufficiently to exploit the advantages of this material. Again once the basic hurdles to recycling have been removed the widespread use of polymer panels could materialize on a wider scale, and continuing production of the GM Saturn derivatives indicate this is possible. If this technology were to be applied now, a complete skin could be envisaged in plastic (SMC or thermoplastic) which could be fully painted using conducting primer, plus intermediate, base and clear coats. Assuming accurate colour matching the base structure might only need an electroprimer thus saving considerable in-house facility and processing costs.

Option 2 must now include the aluminium spaceframe as demonstrated by the Audi A2 and further manufacturers within the premium car sector will be tempted to adopt this strategy, first to counteract the effect of larger engines on the fleet consumption figures and second as a way of assessing the secondary technologies required to mass produce bodies in this material. This would pave the way for volume production if price stability and manufacturing economics became more attractive and the need to drastically lighten bodies became a necessity. As well as creating a new knowledge base to design and manufacture in numbers with sheet, castings and profiles for the substructure, developments within the next 10 years should see a further generation of dent resistant 6xxx series sheet alloy strip for skin panels. For maximum consistency of quality and processing this will take the form of EDT finished, Zr–Ti pretreated strip available in the prelubricated form to minimize any handling damage and surface fretting during transit. As a third option within this high volume category there is no reason why hybrid structures of all three basic materials should not be used with increasing amounts of aluminium and plastics being incorporated in the overall structure over the next 20 years. This would enable the safety and economics associated with steel to be combined with the weight reduction accompanying aluminium, by then a more attractive proposition for closures, and the 'friendliness' and minor

Table 9.1 Possible trends for future materials utilization depending on volume and circumstances (existing examples of the application referred to are shown in brackets)

Throughput		Anticipated conditions			'Accelerated' conditions		
High volume Ca 250 000 p yr.	Main structure	**Option 1** Continued Use of Zinc coated steel — HSS 70% + — Forming grades	**Option 2** Aluminium Spaceframe (Audi A2)	**Option 3** Hybrid mix: — steel structure HSS grades BH, IF, D-P TRIP.	**Option 7** Downsized zinc cozted HSS (ULSAB-AVC) or Aluminium Spaceframe (Audi A2) (GM EV1)	**Option 8** Carbon fibre + aluminium profiles (Vanquish) Pre-prepared carbon fibre construction (Ultima)	**Option 9** Aluminium/composite honeycombe Based platforms
	Closures	Zinc coated BH steel (BMW) or polymer (Saturn)	Aluminium 6016 (Audi A2/A8)	Aluminium 6016 Audi A2/A8) or polymer for bolt-on panels (Saturn)	Polymer panels Polycarbonate Self coloured — (Smart) or (GM EV1) — RIM Horiz. — SMC Vertical	Polymer shell and closures (Elise)	Polymer panels (Smart) self coloured (Saturn)
Low volume	Main structure	**Option 4** Zinc coated steel (L. Rover) (Espace) (SMART)	**Option 5** Aluminium Spaceframe (Ferrari Modena) or punt (Elise) or Monocoque (Jaguar XJ)	**Option 6** Carbon fibre Composite + Al profiles (A-M Vanquish)	**Option 10** Aluminium Spaceframe (Ferrari Modena) or punt (Elise) or Monocoque (Jaguar XJ)	**Option 11** Carbon fibre Composite + Al profiles (A-M Vanquish)	**Option 12** Complete carbon fibre Composite Structure (McLaren F1)
	Closures	Aluminium (L. Rover) or polymer (Saturn) (Espace) (SMART)	Polymer bodyshell (Elise)	Polymer shell or Aluminium	Polymer bodyshell (Elise)	Polymer shell or Aluminium	Polymer Bolt-on panels. (Saturn, SMART)

Table 9.2 Predominant future technologies likely under normal and exceptional circumstances (column width indicative of proportions). Arrows indicate the growing influence of lighterweight options with time

Prevailing circumstances	Volume production		Low/niche car production	
As normal, with anticipated legislation	HS steel structure + BH closures Increasing: • adhesive bonding • laser welding • hydroforms ←	A S F	Sports cars: ASVT + polymer skin ← 4 × 4: hybrid steel + alloy skin panels. Some hydroformed sections	C fibre composite over steel or ASVT frame
Short-term change to meet sudden, more severe fuel conservation or emission controls requirements	Steel hybrid: downsized HSS structure + polymer or aluminium Bolt-'on' parts ←	ASF Aluminium spaceframe • sheet • extrusions • castings Aluminium sheet skin panels	ASVT or hybrid aluminium and polymer skin ←	C fibre composite shell over aluminium honeycomb base + aluminium profiles + composite skin panels

damage tolerance of plastics, for which a solution may then have been found for the recycling of preferred types. Shocks to manufacturing systems may also have been cushioned by intermediate exercises on specific models thus arriving at an optimized cost-effective and environmentally acceptable solution.

Should external factors suddenly accelerate the need for improved fuel consumption/ CO_2 emissions the position would change drastically (if, say, a 3-year response was required resulting from a Middle East conflict). Assuming step changes to power train technology, downsized high strength steel structures using the lightest possible gauges, as shown by ULSAB-AVC, might still be favoured by countries retaining oil resources, e.g. USA, but more pressure would be placed on other areas for ultra-lightweight body structures as the first choice. The call might then be for downsized versions of a simple aluminium spaceframe following the Audi A2 design, for which the production techniques already exist, clad with polymers, i.e. an aluminium-based 'Smart' type of unit, or that of the GM EV1 electric car[11] using adhesively bonded pressings. The still heavyweight power unit would ensure the low centre of gravity needed to provide vehicle stability. Providing lessons were being learnt on mass production techniques that could be applied to honeycomb materials a further five years may then see larger people carrier vehicles based on these materials, and clad with polymer closures. The third choice would assume a step change in carbon fibre composite economics together with manufacturing techniques geared up to volume production and possibly introducing SPS pre-prepared carbon fibre sheet. This would allow the designer to strengthen the structure selectively while embodying the impact resistance already demonstrated in competition cars. Self-coloured panels utilizing gel-coat or similar in-mould techniques would enable major economies to be made in finishing processes and both cosmetic and perforation aspects of corrosion resistance would be vastly improved.

In lower production volumes and including 4×4 structures zinc coated high strength steel still provides the economy and rugged durability required of such structures and this extends to people carriers which may be polymer clad (Espace). The Smart Car illustrates that this combination can be successfully applied to the smaller car. In the niche and sports car markets the aluminium spaceframe (Ferrari Modena) or punt clad with polymer bodyshell (Lotus Elise) will remain popular until the third choice which comprises a carbon fibre composite/aluminium profile substructure plus aluminium skin panels (Aston Martin Vanquish), or predominantly carbon composite structures (McLaren F1, Ultima) prove themselves in extended production and service circumstances. Here, if a short-term change was required this may not prove as drastic a change as for volume cars, as technology including tubular spaceframe, honeycomb and selectively strengthened carbon composite substructures plus polymer skins is already in place. Presumably appropriate power units would be developed to satisfy economy and enhance performance but the main change in bodywork would entail an accelerated move to the right within the material choices indicated in combinations 10–12.

Ideally from a design point of view the Formula 1 technique of optimizing strength where required ensures maximum efficiency of material utilization. Combining feedback from CAD engineering programmes with profile characteristics enables local strengthening exactly where required. A proportion of the manual skill and effort associated with this operation can now be replaced by the use of pre-prepared sheet as demonstrated by SPRINT (SP Resin Infusion Technology) with the Ultima model, the expectation being 'large scale use of composite components in a standard road car within two years'.[12] The impact and endurance properties of carbon fibre composite construction have been proven in F1 competition but again to extrapolate this type of technology to mass produced models even if the materials were economically viable would require major changes. However, this technology may now merit inclusion as an alternative choice within Option 12. A convincing case has been made by reference to the Hypercar paper[13] but in reality the resources needed to influence forward design programmes, rethink supply strategies, networks and economics together with a need to reinvest in totally new facilities, retrain a complete workforce, and change the momentum of large corporations is realistically unlikely in the foreseeable future, although this could happen within a very extended timescale. Should there be sudden shift in emphasis and a need to react within, say, a 3-year timeframe ultralight structures then come into focus as alternative energy sources such as electric systems would require more drastic changes in material type and despite a lack of versatility, sandwich and honeycomb materials are feasible,[14] with simplified designs to facilitate forming and joining.

However, the wheel could yet go full circle, the political and organizational inertia allowing time for the advantages of existing materials to be exploited afresh. It has been assumed that alternative fuel systems will be heavier requiring compensation by the use of ultralightweight body structures. If this were not the case and time allowed the development of compact systems, producing economic fuels and ultra-low emissions, then the pendulum will again swing back in favour of proven, safer, more easily manufactured and recycled steel.

It is emphasized the table indicates an extrapolation of trends evident from current technology and how this may be applied under prevailing conditions, and possibly how this may alter to meet short- to mid-term changes. This model acknowledges the attractiveness of aluminium and composite structures in meeting fuel shortage and emissions legislation stituations likely to arise within the next 10–20 years, but these might only be necessary as stop-gap solutions. The essential factor is the speed with which alternative fuel sources can be introduced and associated economics. The GM fly-by-wire concept Hy-wire vehicle provides timely confirmation of the practicality of fuel cell propulsion and assuming reasonable progress in reducing weight/size of such units and solutions found to safety and cost issues, this radically different type of vehicle power seems a realistic proposition within a 30–40 year timeframe. As the fuel cell mode becomes gradually more dominant the reliance on lightweight construction to promote leaner/cleaner emissions diminishes and reliance on these materials, with inherent manufacturing problems, becomes less. This is not necessarily true of the sports or competitive car where power to weight ratio will always be a primary requirement. Therefore following the excursion into the territory of aluminium/hybrid construction, which has already been witnessed, the fuel cell could then herald a resurgence to cheaper steel structures for which the processing rules are well known and which, at least in America, the perception exists of an inherently safer design material. The predominant structure could become that of an MPV type vehicle with a steel substructure clad with a polymer skin.

9.4 Learning points from Chapter 9

1. Due to the conservative nature of change within the volume car industry radical substitutions in the materials used for body structures are unlikely in the foreseeable future. However, significant progress has been made in the development of steels and aluminium alloys, allowing major weight reductions to be made in mass produced structures fabricated from both these materials.

2. An extensive range of material types has been considered, including proven contenders from other industries, e.g. aerospace, but unless significant cost reductions are evident, and mass production technology demonstrated, these are unlikely to find a wider application than niche or sports car production.

3. The urgency associated with environmental issues such as emissions and landfill controls is not yet of sufficient magnitude as to bring about any more sudden changes. Steel and aluminium present no major recycling problem although plastics still require considerable development into innovative reuse techniques and product rationalization. Existing hybrid structures and planned fuel systems will allow most legislative fuel economy and emissions control legislation to be met in the longer term.

4. Most of the major motor manufacturers have adopted 'in-house' design procedures which identify materials for ease of disassembly/segregation by dismantlers, and also ensure that materials are specified on the basis of known life cycle history regarding selection factors such as energy and costs. This would normally embrace material production, conversion and recycling stages.

5. Competition between the major motor corporations will require the use of designs on an increasingly global basis, to achieve economy of scale. This will call for increasing commonization of platforms, and probably with 'tried and tested' materials. Further engineering initiatives, utilizing increasing computer power, should enable the potential of newer materials to be realized on a longer-term basis, although considerable scope exists for improvement of the uniformity of databases employed, and parameters used, for different materials.

6. Material choices have been made assuming normal circumstances prevail for both volume and niche car conditions. The changes introduced by a sudden 'accelerated' change in conditions such as introduced by a worldwide fuel shortage requiring exceptional fuel economy are also considered. Under normal conditions the steel substructure would continue until the cost of aluminium became more competitive

within the volume production environment. The need for the additional fuel economy under more demanding conditions imposed within a short timeframe would then make aluminium the primary choice – the basic manufacturing rules having already been established for the Audi A2 model. Assuming that progress had been made on the disposal of plastics, it is probable that suitable polymer types could be used to clad the aluminium substructure, dependent on horizontal or vertical orientation.

7. It is likely that niche cars will adopt increasingly hybrid body construction with carbon fibre composite utilized to exploit advantages with regard to safety and weight, once faster manufacturing methods evolve. This trend would be accelerated under conditions of extreme fuel economy where design (minimal weight with strength exactly where required) and power to weight ratio would be enhanced.

8. Materials development is likely to take the form of enhancement of existing materials, e.g. prepainted sheet, 'in-mould coated' polymer panels, alloy modifications, rather than the development and introduction of radically different metals/polymers.

References

1. Webster, B., 'Comparison of Hybrid and Hydrogen Systems', *The Times*, 22 April 2002, p. 8.
2. Gick, M., 'End of Life Vehicles – What Next?', ACP 2000 Automotive Composites and Plastics Conf., Ford Motor Co., Essex, 5–6 Dec. 2000, pp. 211–218.
3. *AEI Magazine*, May 2001, pp. 79–83.
4. Berger, V., 'Concept of Corrosion Protection in the Automotive Industry by Using Organic Coatings over Galvanized Steel Sheet', Galvanized Steel Sheet Forum – Automotive, 15–16 May 2000, Institute of Materials, London, p. 114.
5. Davies, G., 'Vorverdeltes Stahlblech in der Automobilindustrie', *JOT* 1993, heft 9, p. 59.
6. BRITE EURAM 111 Project BE96-3073, 'PVD – Alternatives to Conventional Zinc Coatings'.
7. Maeda, M. *et al.*,'The Product Made by a Vapor Deposition Line', SAE Paper 860273, 1986, p. 179.
8. Duffy, L., 'Magnesium Alloys: The Light Choice for Aerospace', *Materials World,* March 1996, pp. 127–130.
9. North, D., 'Superplastic Alloy for Autobody Construction', British Deep Drawing Research Group Colloquium on Heat Treatment and Metallurgy in Metal Forming, Univ. of Aston, 20 March 1969, Paper No. 5.
10. Clarkson, J., 'And They Still Get it Wrong', *The Sunday Times 'Good Car Bad Car' Guide*, London, 10 Feb. 2002, p. 4.
11. Kobe, G., 'GM EV-1', *Automotive Industries*, Oct. 1996, pp. 69–71.
12. Thomas, S.M., 'Automotive Infusiasm for Composites', *Materials World*, May 2001, p. 20.
13. Fox, J.W. and Cramer, D.R., 'Hypercars: A Market Oriented Approach to Meeting Lifecycle Environmental Goals', SAE Publication SP-1263, Feb. 1997, pp. 163–170.
14. Fenton, J. and Hodkinson, R., *Lightweight Electric/Hybrid Vehicle Design*, Butterworth-Heinemann, pp. 205–206.
15. ULSAB-AVC Report 2002 available from Internet site: www.ulsab-avc.org.

Index